NUTRITION, EPIGENETICS AND HEALTH

NUTRITION, EPIGENETICS AND HEALTH

editors

Graham Burdge
Karen Lillycrop

University of Southampton, UK

World Scientific

NEW JERSEY · LONDON · SINGAPORE · BEIJING · SHANGHAI · HONG KONG · TAIPEI · CHENNAI · TOKYO

Published by

World Scientific Publishing Co. Pte. Ltd.
5 Toh Tuck Link, Singapore 596224
USA office: 27 Warren Street, Suite 401-402, Hackensack, NJ 07601
UK office: 57 Shelton Street, Covent Garden, London WC2H 9HE

Library of Congress Cataloging-in-Publication Data
Names: Burdge, Graham, editor. | Lillycrop, Karen, editor.
Title: Nutrition, epigenetics and health / edited by Graham Burdge & Karen Lillycrop.
Description: New Jersey : World Scientific, 2016. | Includes bibliographical references.
Identifiers: LCCN 2016025873 | ISBN 9789813143302 (hardcover : alk. paper)
Subjects: | MESH: Nutrigenomics | Epigenomics--methods | Genetic Processes | Risk Factors
Classification: LCC RB155 | NLM QU 145 | DDC 616/.042--dc23
LC record available at https://lccn.loc.gov/2016025873

British Library Cataloguing-in-Publication Data
A catalogue record for this book is available from the British Library.

Copyright © 2017 by World Scientific Publishing Co. Pte. Ltd.

All rights reserved. This book, or parts thereof, may not be reproduced in any form or by any means, electronic or mechanical, including photocopying, recording or any information storage and retrieval system now known or to be invented, without written permission from the publisher.

For photocopying of material in this volume, please pay a copying fee through the Copyright Clearance Center, Inc., 222 Rosewood Drive, Danvers, MA 01923, USA. In this case permission to photocopy is not required from the publisher.

Typeset by Stallion Press
Email: enquiries@stallionpress.com

Printed in Singapore

Contents

Preface		vii
Chapter 1	Basic Mechanisms in Epigenetics Karen A. Lillycrop	1
Chapter 2	Nutrition, Epigenetics and the Early Life Origins of Disease: Evidence from Human Studies Nina D'Vaz and Rae-Chi Huang	25
Chapter 3	The Early Life Nutritional Environment, Epigenetics and Developmental Programming of Disease: Evidence from Animal Models Mark H. Vickers	41
Chapter 4	Lipids and Epigenetics Graham C. Burdge	73
Chapter 5	Circadian Biology: Interaction with Metabolism and Nutrition Jonathan D. Johnston	93
Chapter 6	Nutrition, Epigenetics and Aging John C. Mathers and Hyang-Min Byun	103
Chapter 7	Nutrition and Epigenetics: Evidence for Multi- and Transgenerational Effects Cheryl S. Rosenfeld	133
Chapter 8	Epigenetic Biomarkers and Global Health Paula Costello and Mark Hanson	159

| Chapter 9 | Nutrition, Epigenetics and Health: Evolutionary Perspectives
Sinead English and Tobias Uller | 177 |
| Chapter 10 | The Body Politic: Epigenetics and Society
Noela Davis | 201 |

Index 221

Preface

What and how much we eat has an important influence on our health and our ability to function as humans. Research over the past hundred years or so has provided substantial information about nutritional requirements and the consequences of these not being met by the diet. These have led to the formulation dietary recommendations to populations, with some limited adjustment for age, sex and pregnancy, in order to promote health. However, this approach does not account for individual variation within these population sub-groups. The importance of such differences in nutritional requirements has been highlighted in recent years by the development of the concept of personalised nutrition based on nutrigenomics. However, variation in phenotype between individuals is not captured completely by fixed mutations in the genome. Recent rapid developments in the science of epigenetics have demonstrated that such processes are an important source of individual variation and contribute substantially to risk of diseases of global importance including obesity, cardiovascular disease and cancer. One implication of these studies is that unlike gene polymorphisms, epigenetic processes can be modified by environmental factors. This suggests an opportunity to develop personalised nutrition strategies that to correct epigenetic marks that predispose to disease. Unfortunately, at the time of writing such interventions are some distance into the future as there are a substantial number of challenges to be overcome not least of which are the potential for epigenetic change across the life course and across generations. Our ambition for this book was to capture something of range and potential of the science of nutriepigenomics by bringing together the thoughts of leading researchers in this field and to set these in the context of global health policy and the ethical implications for society. We hope that in reading this book you will gain something of the excitement of this emerging field.

Karen Lillycrop (PhD)
Professor of Epigenetics
Faculty of Natural and Environmental Sciences
University of Southampton
Southampton
UK

Graham Burdge (PhD)
Professor of Nutritional Biochemistry
Faculty of Medicine
University of Southampton
Southampton
UK

Basic Mechanisms in Epigenetics

CHAPTER 1

Karen A. Lillycrop

Centre for Biological Sciences, University of Southampton, IDS Building (MP 887), Southampton General Hospital, Tremona Road, Southampton, SO16 6YD, UK

Abbreviations

5mC, 5 methyl cytosine;
5hmC, 5 hydroxy methyl cytosine;
5fC, 5-formylcytosine;
5caC, 5-carboxylcytosine;
CBX, chromobox-domain protein;
C/EBPα, CCAAT/Enhancer Binding Protein;
CRE, cAMP response element;
CpG, cytosine and guanine nucleotides linked by phosphate;
CTCF, CCCTC-Binding Factor (Zinc Finger Protein);
Dnmt, DNA methyltransferase;
EED, embryonic ectoderm development;
ES, embryonic stem;
EZH2, enhancer of zeste homolog 2;
HATs, histone acetyl transferases;
HDACs, histone deacetylases;
HMTs, histone methyl transferases;
JMJD2, Jumonji domain 2;
IAP, intracisternal A-particles;
ICR, Imprinting control region;
IGF-2, insulin like growth factor-2;
LINE, long interspersed nuclear elements;

lncRNAs, long non-coding RNAs;
LSD, lysine-specific demethylase 1;
LTR, Long terminal repeats containing endogenous retroviruses;
MBD, methyl binding domain;
MeCP2, methyl CpG binding protein 2;
miRNAs, microRNAs;
ncRNAs, non-coding RNA;
NRF-1, Nuclear respiratory factor 1;
PARP, poly-ADP-ribose polymerases;
piRNA, Piwi-interacting RNA;
PGC1α, peroxisome proliferator- activated receptor gamma, coactivator 1 alpha;
PRC, polycomb repressive complex;
PPAR, peroxisomal proliferator-activated receptor;
O-GlcNAc, β-N-acetylglucosamine;
O-GlcNAcase, β-N-acetylglucosaminidase;
SAM, s adenosyl methionine;
SINE, short interspersed nuclear elements;
SIRT1, Sirtuin1;
SUZ12, suppressor of zeste 12;
RISC, RNA-induced silencing complex;
TET, Ten-eleven translocation proteins;
YY1, Yin Yang 1;
XCI, X chromosome inactivation;
Xist, X-inactive specific transcript

Introduction

It has been widely recognized that the sequence of bases (A,C,G,T) within DNA determine phenotype. However, recent research has shown that gene sequence is not the sole determinant of phenotype, but epigenetic processes can also influence phenotypic traits. Epigenetic processes regulate the accessibility of genes to the cellular proteins that regulate gene transcription, determining where and when a gene is switched on, and its level of activity. Epigenetic processes not only play a central role in regulating gene expression but also allow an organism to adapt to the environment. A growing body of evidence has shown that a number of environmental factors such as nutrition, body composition or social environment can all influence these epigenetic processes, often with long-term consequences on metabolism and disease risk. In this first chapter, the basic mechanisms of the three main epigenetic processes, DNA methylation, histone modification and non-coding RNAs will be discussed.

Introduction to Epigenetics

Conrad Waddington (1905–1975) first introduced the term "epigenetics" in the 1940s. He used it to define "the branch of biology which studies the causal interactions between genes and their products, which bring the phenotype into being".[1] However, with our increased understanding of the mechanisms underlying gene regulation and cell specification, the definition of epigenetics has narrowed and in 2007, Adrian Bird defined epigenetics as "the structural adaptation of chromosomal regions so as to register, signal or perpetuate altered activity states". But there was much controversy over how stable the changes induced by such epigenetics processes needed to be to fit the definition of epigenetic and so in 2008, a consensus definition was produced and epigenetics was defined as "stably heritable phenotypes resulting from changes in a chromosome without changes in gene sequence".[2] Epigenetic processes include: DNA methylation, histone modifications, and non-coding RNAs. These processes regulate all aspects of gene expression by controlling access to the underlying DNA sequence. Epigenetic mechanisms also provide both variability and rapid adaptability that allow organisms to respond to the environment both in the short and long term.

DNA Methylation

DNA methylation is a common modification in eukaryotic organisms. In mammals, DNA methylation occurs primarily on the 5th

position of the cytosine (5mC) base within a CpG dinucleotide. In humans, this dinucleotide is present at only 5 to 10% of its predicted frequency with a total 28 million CpGs within the human genome, of which 70 to 80% are methylated.[3] These CpG dinucleotides are not uniformly distributed across the genome but are clustered at the 5' promoter or control regions of genes. This cluster of CpGs is often referred to as a CpG island. CpG islands are defined as a sequence with a G+C content of greater than 60% and ratio of CpG to GpC of at least 0.6.[4] Genes which contain CpG islands are usually "housekeeping" genes which are constitutively expressed and where the CpG islands are largely unmethylated. Methylated CpG islands are however found in the promoters of developmental genes[5,6] suggesting that DNA methylation is an important regulatory mechanism in regulating cell fate. Other sequences that are heavily methylated in mammals includes repetitive DNA sequences such as those within the centromeric and pericentromeric regions of chromosomes or in the endogenous transposable elements such as the long interspersed nuclear elements (LINEs), short interspersed nuclear elements (SINEs) and long terminal repeat (LTR)-containing endogenous retroviruses.[3]

DNA Methylation Regulates Gene Transcription

The observation that low levels of DNA methylation in the promoter or control regions of genes are generally associated with transcriptional activity, while high levels of promoter methylation are associated with transcriptional silencing, led to the suggestion that DNA methylation may modulate the level of gene transcription.[3] Subsequent studies support this hypothesis and have shown that DNA methylation can affect gene transcription through two main mechanisms. Firstly, DNA methylation can influence transcription factor binding. In the vast majority of cases, DNA methylation within the binding site of a transcription factor has been shown to block binding of the transcription factor to the DNA. Examples of transcription factors whose binding is inhibited by

DNA methylation include NRF-1,[7] YY1, MYC and AP1.[8] But there are exceptions, for example, methylation of the CRE sequence enhances the DNA binding of C/EBPα, which in turn activates a set of promoters specific for adipocyte differentiation.[9,10]

The second mechanism by which DNA methylation can affect gene transcription is through the recruitment of histone modifying enzyme complexes to the DNA which leads to transcriptional repression.[11] Methyl CpG binding protein 1 (MeCP1) recognizes and binds to methylated CpGs but requires at least seven methylated CpGs for efficient DNA binding and so recruitment of MeCP1 may be important for the repression of genes with CpG islands,[12,13] while MeCP2 binds to a single methylated CpG and may be responsible for repression from CpG sparse promoters.[14]

However, DNA methylation is also found within the bodies of genes, and here higher levels of intragenic methylation correlate with higher levels of gene expression.[15–17] The functional significance of gene body methylation is less clear and a number of potential roles have been suggested, including regulation of alternative transcription initiation sites[17] or regulation of splicing.[18] DNA methylation in mammals is often increased at exons relative to introns, leading to the suggestion that DNA methylation may mark exon/intron boundaries.[19] It has been proposed that DNA methylation may modulate the binding of DNA-binding factors such as CTCF, which could induce a local pausing of the RNA polymerase and favor the co-transcriptional assembly of the spliceosome at splice sites.[20]

Non-CpG Methylation

In mammals, non-CpG methylation has also been reported. This occurs primarily at CpA dinucleotides and is most prevalent in embryonic stem (ES) cells, oocytes and in neuronal cells in the brain. Non-CpG methylation is enriched at certain genomic features and co-occurs with high levels of CpG methylation.[21] Interestingly it has been shown that non-CpG methylation is drastically reduced upon ES cell differentiation but the functional consequences of this are not known. Non-CpG methylation has also been recently reported within the promoter of the PGC1α gene, which is involved in insulin secretion, in islet cells of patients with T2D compared to controls,[22] suggesting that non-CpG methylation may play a role in the dysregulation of PGC1α and development of this disease.

DNA Methylation can be Regulated Dynamically

In mammals, DNA methylation is catalyzed by the DNA methyltransferases (DNMTs). There are three main DNMTs. DNMT1 is the maintenance DNMT which copies the methylation mark from the parental strand of DNA to the newly synthesized strand during the process of DNA replication, while DNMT3A and DNMT3B are responsible for establishing *de novo* DNA methylation.[13]

DNA methylation was originally thought to be a very stable modification which once established was then maintained throughout the life course of the organism. This premise was based on the thermodynamic stability of the methylation mark and the initial uncertainty of whether a biochemical mechanism existed by which the methyl group could be directly removed from 5mC.[17] Thus it was thought that DNA demethylation could only occur passively through a failure to maintain DNA methylation levels after the process of DNA replication. However, this concept has now been shown to be incorrect as there is growing evidence that active demethylation does occur. For example, active demethylation in mammals has been observed on the paternal genomic DNA in the zygote upon fertilization,[18,19] on the synaptic plasticity gene reeling in the hippocampus upon contextual fear conditioning[19,20] and on the interferon gamma gene upon antigen exposure of memory CD8 T cells.[20] Moreover, a number of potential DNA demethylases have been discovered including Ten-eleven translocation (TET) proteins,[21,22] methyl binding domain protein (MBD)2b,[23] MBD4,[24] the DNA repair endonucleases XPG (Gadd45a)[25] and a G/T mismatch repair DNA glycosylase.[26] These potential DNA demethylases function not by removing the methyl group from cytosine directly, but through a multistep process linked either to DNA repair mechanisms or through further modification of 5mC.

TET Proteins and DNA Hydroxymethylation

In 2009, 5-hydroxymethylcytosine (5hmC) was discovered as another cytosine modification. Recent studies have identified the Tet proteins, a family of Fe(II) and α-ketoglutarate-dependent dioxygenases, as the enzymes responsible for catalyzing the conversion of 5mC to 5hmC.[23] The mammalian TET family contains three members, TET1, TET2 and TET3, all of which share a high degree of homology within their C-terminal catalytic domain.[24,25] Furthermore, recent findings have shown that TET proteins can oxidize 5mC not only to 5hmC but also to 5-formylcytosine (5fC) and/or 5-carboxylcytosine (5caC),[26,27] which are subsequently excised by thymine DNA glycosylase,[28,29] or deaminated by activation-induced deaminase, whose deamination product, 5-hydroxymethyluracil, activates base-excision repair pathway leading to demethylation.[30]

The level of 5hmC varies during development. High levels of 5hmC are found in ES cells, in multipotent adult stem cells and progenitor cells. But upon differentiation, levels decrease,[23,31] although in Purkinje neurons and other neural subtypes, levels remain high.[32] However, the relative stability of 5hmC, raises the question as to whether it is simply an intermediate in the demethylation of DNA or whether 5hmC may also serve as an epigenetic mark with its own regulatory role. The distribution of 5hmC, like 5mC is not uniform across the genome, while 5mC is enriched in pericentromeric heterochromatin, 5hmC is associated euchromatin, supporting the premise that 5hmC is associated with gene activation. Recent genome-wide sequencing studies have also shown that in ES cells, both TET1 and 5hmC are enriched within gene bodies and at transcription start sites (TSSs) and promoters.[33–35] The presence of 5hmC in gene bodies appears to be associated with gene expression[33,34,36] and it has been suggested that 5hmC may relieve the silencing effect of 5mC by preventing binding of methyl-binding proteins.[37] There is less of a clear association between enrichment of 5hmC at promoters and gene activation. Interestingly after TET1 knockdown in ES cells, although several genes are downregulated, many genes are upregulated.[33,38,39] Many of these were developmental control genes containing bivalent gene promoters, which are characterized by histone modifications associated with both gene activation (trimethylated histone H3 lysine 4-H3K4me3) and repression (trimethylated histone H3 lysine 27-H3K27me3).[33,34] Both TET1 and 5hmC are enriched at bivalent genes, and TET1 depletion has been shown to impair binding of the polycomb complex which is responsible to H3K27 methylation and gene repression, suggesting that maintenance of hypomethylation by TET1 allows PRC2 binding.[34,35,38,39] Such studies imply TET1 and 5hmC may have multiple regulatory roles both in the demethylation of DNA and in regulating histone modification.

Establishment of DNA Methylation Patterns through Development

DNA methylation is markedly reprogrammed during embryonic development in mammals. Following fertilization, the methylation marks on the maternal and paternal genomes are largely erased.[40] Immunofluorescence studies have shown that this global demethylation occurs asymmetrically. Firstly, demethylation occurs rapidly on the paternal DNA, catalyzed by the TET proteins, while in contrast demethylation on the maternal DNA occurs more slowly over several cell divisions by a replication-dependent passive process. Recent bisulfite sequencing studies have confirmed that blastocysts contain a demethylated genome.[41–43] A number of regions escape this erasure and these include the imprinting control regions (ICRs) of imprinted genes and certain classes of mobile elements such as intracisternal A-particles (IAPs).[44] After this phase of global demethylation and around the time of implantation, the genome then undergoes *de novo* methylation. The pluripotency genes *Oct-3/4* and *Nanog*, which are essential to maintain the undifferentiated state of the early stem cells are methylated and silenced.[45] Lineage-specific methylation of tissue-specific genes also occurs throughout prenatal development and early postnatal life which is essential for cell specification.[46] There is also methylation of the repetitive DNA sequences, endogenous viral sequences and many cryptic promoters at this time. In contrast, CpG islands within housekeeping genes are protected from this global de novo methylation and remain unmethylated.[40]

Histone Modifications

In eukaryotic organisms, the DNA within the nucleus is wrapped around a core of eight histone proteins (H2A, H2B, H3 and H4) forming the basic unit of chromatin, a nucleosome. Each nucleosome is then folded upon itself to form a solenoid or 30nm fiber which is then further coiled and compacted to form a 200nm fiber. This folding and packaging of the DNA is essential to reduce its effective size. For example, the total extended length of DNA in a human cell is nearly 2m, yet DNA fits into the nucleus of a cell with a diameter of only 5 to 10 μm.[47,48] However, it is now clear that histone proteins are not only important for the packaging of DNA but also play a critical role in regulating gene expression alongside DNA methylation. The histone proteins contain two domains; a globular domain and an amino tail domain. The amino tail domain of the histone proteins are accessible, unstructured domains that protrude out of the nucleosomes. These tail domains are rich in positively charged amino acids which interact with the negatively charged DNA and are now known to be subject to a large number of post-translational including acetylation, methylation, phosphorylation; ubiquitination; sumoylation; ADP ribosylation; glycosylation; biotinylation, and carbonylation.[49,50] The establishment of these marks on the histone tails is often referred to as the histone code which states that "post translational modification of specific amino acid residues within the histone leads to the binding of effector proteins that, in turn, bring about specific cellular processes".[51]

Histone Acetylation

Allfrey first reported histone acetylation back in 1964.[52] Histone acetylation is a highly dynamic process which is regulated by histone acetyltransferases (HATs) and histone deacetylases (HDACs). The HATs utilize acetyl CoA as a cofactor and catalyze the transfer of an acetyl group to the ε-amino group of the lysine (K) side chains.[49] This neutralizes the positive charge on the lysine, weakening the interaction between the histones and DNA. In contrast, the HDAC enzymes remove the acetyl group from the lysines in the histone tails, restoring the positive charge of the tail leading to a tightening of the chromatin structure and gene repression.

Histone Phosphorylation

Histone phosphorylation occurs on serines, threonines and tyrosines mainly in the amino terminal tails of histones. Histone phosphorylation is catalyzed by histone kinases which transfer a phosphate group from ATP to the hydroxyl group of the amino-acid side chain, adding negative charge to the histone tail. This modification is generally associated with gene activation. Histone phosphatases remove the modification.[53]

Histone Methylation

Histone methylation occurs mainly on lysines and arginines and can be either an active or repressive mark depending on the specific residue involved.[30] For example, histone H3K4, H3K36, H3K79 methylation is associated with gene activation, while histone H3K9, H3K27, H4K20 methylation is linked to gene silencing.[31] Furthermore, lysines may be mono-, di- or tri-methylated, whereas arginines may be mono-, symmetrically or asymmetrically di-methylated.[54] The first histone lysine methyltransferase to be identified was suppressor of variegation 3-9 homolog 1 (SUV39H1) which targets H3K9.[55] Numerous histone methyl transferases have since been identified, they all catalyze the transfer of a methyl group from S-adenosylmethionine (SAM) to a lysine's ε-amino group.[56]

The polycomb repressive complex (PRC1 and PRC2) mediates H3K27 trimethylation (me3) which is a repressive histone mark. The PRC2 complex which contains three main proteins — enhancer of zeste 2 (EZH2) or its close homolog EZH1, embryonic ectoderm development (EED), and suppressor of zeste 12 (SUZ12) induces H3K27 methylation.[57,58] H3K27me3 then acts as a docking site for the chromobox-domain (CBX) protein subunits of PRC1, allowing PRC1 to bind which is important for the maintenance of the histone mark.[59] Polycomb proteins and H3K27 are found associated with silenced genes, although while PRC2 is enriched at CpG islands close to transcriptional start sites, H3K27me3 and PRC1 cover parts of the gene bodies as well. In pluripotent ES cells, H3K27me3 is often found at developmentally regulated genes together with the active histone H3K4me3 mark. The presence of both active and repressive marks on a gene is referred to as a bivalent domain and is associated with the poised expression of developmental control genes which may allow their activation in the presence of differentiation signals while maintaining their expression in the absence of signals.[60,61] Consistent with the importance of this mark in regulating developmental control genes, many components of the PRC1/2 complexes have been found to be overexpressed in human cancers[62] and interestingly many of the polycomb target genes are hypermethylated in cancer and aging.[63]

Like DNA methylation, for many years, histone methylation was considered a stable modification, however a number of histone demethylases have now been described. For instance, lysine-specific demethylase 1 (LSD1) removes methyl groups from H3K4me1/2,[64] while JMJD2 is a tri-methyl lysine demethylase that demethylates H3K9me3 and H3K36me3.[65]

Other histone modifications include the addition of β-N-acetylglucosamine (O-GlcNAc) sugar residues to H2A, H2B and H4. This modification, like the other histone marks, is highly dynamic with O-GlcNAc transferase, catalyzing the transfer of the sugar to the histone and β-N-acetylglucosaminidase (O-GlcNAcase) removing the sugar residue.[66] Histones can also be mono- and poly-ADP ribosylated on glutamate and arginine residues, and this is associated with a relaxed open chromatin state.[67] Poly-ADP-ribosylation of histones is catalyzed by the poly-ADP-ribose polymerases (PARP) and reversed by the poly-ADP-ribose-glycohydrolase enzymes.[68,69]

Mode of Action of Histone Modifications

Histone modifications exert their effects via two main mechanisms. The first mechanism involves the histone modification directly influencing the overall chromatin structure. Histone acetylation and phosphorylation reduce the positive charge of the histones, and this weakens the electrostatic interactions between the histones and the negatively charged DNA, leading to a more open chromatin structure facilitating access to the transcriptional machinery. Histone acetylation occurs on numerous lysines in the amino tail of the histone, including H3K9, H3K14, H3K18, H4K5, H4K8 and H4K12,[70] leading to hyper-acetylation of promoter regions and a change in chromatin structure. The second mechanism by which histone modifications may affect gene expression is through the recruitment of effector proteins which bind specifically to the modified histone. Numerous chromatin-associated factors have been shown to specifically interact with modified histones via distinct domains.[70] For instance, trimethylation of H3K9 leads to the recruitment and binding of Heterochromatin protein (HP)1, a protein associated with repressive chromatin. HP1 binds to H3K9me3 via its N-terminal chromodomain and this interaction is important for the maintenance of heterochromatin.[71,72]

Crosstalk between DNA methylation and histone modification is well known. Methylated DNA is bound by MeCP2 which recruits both histone deacetylases, which remove acetyl groups from the histones,[32–34] and histone methyl transferases such as Suv39H1[31] which methylates K9 on histone H3, resulting in a closed chromatin structure and transcriptional silencing. Recent studies have also shown that DNMT1 can be recruited by a number of histone modifying enzymes such as HDAC1 and HDAC2, as well as components of the PRC2,[35,36] suggesting that chromatin structure may also determine DNA methylation status.

Non-coding RNA

Although up to 90% of the eukaryotic genome is transcribed,[37] only 1–2% of the genome encodes proteins.[38] This suggests that a large proportion of the RNA molecules synthesized do not encode proteins. These are termed non-coding RNAs (ncRNAs). NcRNAs can be grouped into two classes according to the size of the transcripts. Long ncRNAs are longer than 200 nucleotides, while short ncRNAs are less than 200 nucleotides.

Short ncRNAs include microRNAs (miRNAs), small interfering RNAs (siRNAs) and Piwi-interacting RNAs (piRNAs).[39] The most studied to date of these small non-coding RNAs are the miRNAs. miRNAs are typically 21–23 nt in length, evolutionary conserved, single-stranded, non-coding RNA molecules. They are synthesized as a precurosr miRNAs (Primary miRNA) from which mature miRNAs are generated through a two-step cleavage process. The mature miRNA is then incorporated into the RNA-induced silencing complex (RISC), here the miRNA functions to guide the RISC complex to the target mRNA. The level of complementarity between the miRNA and mRNA target determines which silencing mechanism will be employed; cleavage of target messenger RNA (mRNA) with subsequent degradation or translation inhibition.[40–42]

In contrast, the mechanism by which long non-coding RNAs work is poorly understood, but current evidence suggests that long ncRNAs can recruit histone modifying complexes to the DNA to create repressive domains covering many kilobases. In mammals, long ncRNAs have been shown to be essential for X-chromosome inactivation and genomic imprinting.[43–46] In humans, the antisense long ncRNA HOTAIR, transcribed from the HOXC locus, regulates gene expression at HOXD locus in trans by recruiting PRC2,[47] which mediates histone H3K27 methylation. HOTAIR physically associates with the PRC2 and modulates PRC2 activity to deposit

H3K27 trimethylation marks at target chromatin throughout the genome to induce gene silencing.[48] HOTAIR has also been shown to be overexpressed in approximately one quarter of human breast cancers, directing PRC2 to approximately 800 ectopic sites in the genome, which leads to histone H3 lysine 27 trimethylation and changes in gene expression.[73] Long-range regulation by lncRNAs may be a widespread mechanism, more than 20% of tested lncRNAs are bound by PRC2 and other chromatin modifiers.[74]

Function of Epigenetic Processes

DNA methylation, histone modification and non-coding RNAs play an essential role in regulating gene expression, controlling tissue specific gene expression and cell differentiation. These epigenetic processes are also responsible for maintaining the asymmetrical silencing of imprinted genes[75] and X chromosome inactivation.[76,77]

Genomic Imprinting

There are about 100 so called imprinted genes within the human genome. These are genes whose expression is restricted by an epigenetic mechanism to either the maternal or paternal allele. For each imprinted gene, it is either the maternal or paternal allele that is expressed, for instance IGF2 is an imprinted gene which regulates fetal growth. It is always the paternally inherited copy that is expressed and the maternally inherited copy repressed. Although DNA methylation was thought to be the primary mechanism of gene silencing and certainly is important for maintaining the transcriptional silencing of the inactive gene, there is emerging evidence that lncRNAs may direct this process via recruitment of histone repressive complexes and DNMTs.[78] Disruption to the epigenetic silencing of the imprinted gene loci has been associated with a number of imprinted disorders including Prader-Willi syndrome, Angelman syndrome, and several types of cancer.[79]

X Inactivation

X inactivation (XCI) is a closely related process that equates levels of gene expression between males and females by inactivating one of the X chromosomes in female cells. However, unlike genomic imprinting, X inactivation is a random process. Xist, a lncRNA plays a key role in XCI.[80] During female development, Xist RNA is expressed from the inactive X, Xist then "coats" the X chromosome from which it is transcribed, leading to chromosome-wide repression of gene expression. An overlapping antisense lncRNA called Tsix then represses Xist expression in *cis*, while the Jpx, another lncRNA activates Xist on the inactive X.[81]

Environmental Modulation of the Epigenome

Although DNA methylation was originally thought to be a very stable modification and once established, methylation patterns were largely maintained, there is now growing evidence that a number of environmental factors such as nutrition,[82] body composition,[83] endocrine disruptors[84] and social environment[85] can all modulate the epigenome, often resulting in long-term changes in gene expression and metabolism. The epigenome seems most susceptible to environmental factors in early life particularly during the prenatal, neonatal and pubertal periods when the epigenome is being established.[86] One factor that has been shown consistently to alter the epigenome is nutrition, with long lasting effects on health. Both human and experimental studies have shown that dietary restriction, protein restriction, folic acid intake, high fat or high glucose diets can all induce persistent changes in DNA methylation, histone modifications and/or miRNA expression. The ability of nutritional factors to influence the epigenome is perhaps not unexpected as many transcription factors which recruit the writers of the epigenetic code are regulated by nutritional factors. For instance, the PPAR family of nuclear receptors which play a key role in lipid metabolism are activated by polyunsaturated fatty acids,[87] while

Sirt1, which acts as an energy sensor regulating many aspects of metabolism is a NAD dependent protein deacetylase.[88]

In addition, many readers and writers of the epigenetic code are themselves regulated at least in part by the concentration of specific metabolic substrates or cofactors. Methyl groups for all biological methylation reactions including DNA and histone methylation are primarily supplied from dietary methyl donors and cofactors via one-carbon metabolism. In this pathway, methionine is converted to S-adenosylmethionine, the universal methyl donor. After transferring the methyl group, S-adenosylmethionine is then converted to S-adenosylhomocysteine, which in turn is converted to homocysteine. Homocysteine is then either recycled to methionine by the enzyme betaine homocysteine methyltransferase, which uses betaine or choline, or via a folate-dependent remethylation pathway, where 5-methyl Tetrahydrofolate is reduced to 5,10-methylene Tetrahydrofolate by 5,10-methyleneTetra-hydrofolate reductase. This methyl group is then used by methionine synthase to convert homocysteine to methionine using vitamin B12 as the cofactor. This reliance on dietary sources of methyl donors and cofactors for DNA methylation reactions is consistent with a growing body of evidence that changes in the supply of methyl donors or cofactors through dietary intake can alter the epigenome and induce long-term changes in gene expression.[89,90]

But it is not just the methyl transferases that may be affected by diet and energy intake. HATs transfer the acetyl group from acetyl CoA to the lysines residues of the histone tails,[91] suggesting that the availability of acetyl-CoA may influence histone acetylation rates. Consistent with this, it has been shown that intracellular acetyl-CoA concentration show a ~10-fold variation[92] and since the K_m of most HATs is within this range, the activities of HATs may be sensitive to variations in intracellular acetyl-CoA levels. LSD1, which demethylates H3K4 uses the reduction of the cofactor flavin adenine dinucleotide (FAD) to $FADH_2$,[93] Jumonji-C domain containing histone demethylases and the TET family of proteins utilizes α-ketoglutarate (αKG),[23,94] both of which are key metabolites of the TCA cycle. Thus, variations in dietary intake in terms of either individual components, or total energy is likely to have an effect on the epigenome and potentially induce a persistent change on gene expression.

Conclusion

The discovery of the epigenetic code, this second code which controls gene activity is having a profound effect on not only our knowledge of basic biological mechanisms of gene regulation but also on how the environment can shape our genome and phenotype. Consistent with the importance of these processes in gene control there is now growing evidence that alterations to these processes especially in early development can affect disease susceptibility. An understanding of the role of epigenetic process in the development of disease will potentially lead to the early identification of individuals at risk and novel intervention strategies to treat or alleviate disease.

References

1. CH Waddington. (1942) Canalization of development and the inheritance of aquired characters. *Nature* **150**:563–565.
2. SL Berger, T Kouzarides, R Shiekhattar, A Shilatifard. (2009) An operational definition of epigenetics. *Genes Dev* **23**:781–783.
3. A Bird. (2002) DNA methylation patterns and epigenetic memory. *Genes and Development* **16**:6–21.
4. SH Cross, AP Bird. (1995) CpG islands and genes. *Curr Opin Genet Dev* **5**:309–314.
5. S Saxonov, P Berg, DL Brutlag. (2006) A genome-wide analysis of CpG dinucleotides in the human genome distinguishes two distinct classes of promoters. *Proc Natl Acad Sci USA* **103**:1412–1417.
6. R Illingworth, A Kerr, D Desousa, H Jorgensen, P Ellis *et al.* (2008) A novel CpG island set identifies tissue-specific methylation at developmental gene loci. *PLoS Biol* **6**:e22.
7. JR Blesa, AA Hegde, J Hernandez-Yago. (2008) In vitro methylation of nuclear respiratory factor-2 binding sites suppresses the promoter activity of the human TOMM70 gene. *Gene* **427**:58–64.

8. PH Tate, AP Bird. (1993) Effects of DNA methylation on DNA-binding proteins and gene expression. *Curr Opin Genet Dev* **3**:226–231.
9. V Rishi, P Bhattacharya, R Chatterjee, J Rozenberg, J Zhao *et al*. (2010) CpG methylation of half-CRE sequences creates C/EBPalpha binding sites that activate some tissue-specific genes. *Proc Natl Acad Sci USA* **107**:20311–20316.
10. R Chatterjee, C Vinson. (2012) CpG methylation recruits sequence specific transcription factors essential for tissue specific gene expression. *Biochim Biophys Acta* **1819**:763–770.
11. F Fuks, PJ Hurd, D Wolf, X Nan, AP Bird *et al*. (2003) The methyl-CpG-binding protein MeCP2 links DNA methylation to histone methylation. *Journal of Biological Chemistry* **278**:4035–4040.
12. RR Meehan, JD Lewis, S McKay, EL Kleiner, AP Bird. (1989) Identification of a mammalian protein that binds specifically to DNA containing methylated CpGs. *Cell* **58**:499–507.
13. JP Jost, J Hofsteenge. (1992) The repressor MDBP-2 is a member of the histone H1 family that binds preferentially in vitro and in vivo to methylated nonspecific DNA sequences. *Proc Natl Acad Sci USA* **89**:9499–9503.
14. HH Ng, P Jeppesen, A Bird. (2000) Active repression of methylated genes by the chromosomal protein MBD1. *Mol Cell Biol* **20**:1394–1406.
15. A Hellman, A Chess. (2007) Gene body-specific methylation on the active X chromosome. *Science* **315**:1141–1143.
16. L Laurent, E Wong, G Li, T Huynh, A Tsirigos *et al*. (2010) Dynamic changes in the human methylome during differentiation. *Genome Res* **20**:320–331.
17. AK Maunakea, RP Nagarajan, M Bilenky, TJ Ballinger, C D'Souza *et al*. (2010) Conserved role of intragenic DNA methylation in regulating alternative promoters. *Nature* **466**:253–257.
18. JI Young, EP Hong, JC Castle, J Crespo-Barreto, AB Bowman *et al*. (2005) Regulation of RNA splicing by the methylation-dependent transcriptional repressor methyl-CpG binding protein 2. *Proc Natl Acad Sci USA* **102**:17551–17558.
19. RF Luco, M Allo, IE Schor, AR Kornblihtt, T Misteli. (2011) Epigenetics in alternative pre-mRNA splicing. *Cell* **144**:16–26.

20. S Shukla, E Kavak, M Gregory, M Imashimizu, B Shutinoski et al. (2011) CTCF-promoted RNA polymerase II pausing links DNA methylation to splicing. *Nature* **479**:74–79.
21. BH Ramsahoye, D Biniszkiewicz, F Lyko, V Clark, AP Bird et al. (2000) Non-CpG methylation is prevalent in embryonic stem cells and may be mediated by DNA methyltransferase 3a. *Proc Natl Acad Sci USA* **97**:5237–5242.
22. R Barres, ME Osler, J Yan, A Rune, T Fritz et al. (2009) Non-CpG methylation of the PGC-1alpha promoter through DNMT3B controls mitochondrial density. *Cell Metab* **10**:189–198.
23. M Tahiliani, KP Koh, Y Shen, WA Pastor, H Bandukwala et al. (2009) Conversion of 5-methylcytosine to 5-hydroxymethylcytosine in mammalian DNA by MLL partner TET1. *Science* **324**:930–935.
24. LM Iyer, M Tahiliani, A Rao, L Aravind. (2009) Prediction of novel families of enzymes involved in oxidative and other complex modifications of bases in nucleic acids. *Cell Cycle* **8**:1698–1710.
25. C Loenarz, CJ Schofield. (2009) Oxygenase catalyzed 5-methylcytosine hydroxylation. *Chem Biol* **16**:580–583.
26. YF He, BZ Li, Z Li, P Liu, Y Wang et al. (2011) TET-mediated formation of 5-carboxylcytosine and its excision by TDG in mammalian DNA. *Science* **333**:1303–1307.
27. S Ito, L Shen, Q Dai, SC Wu, LB Collins et al. (2011) TET proteins can convert 5-methylcytosine to 5-formylcytosine and 5-carboxylcytosine. *Science* **333**:1300–1303.
28. T Pfaffeneder, B Hackner, M Truss, M Munzel, M Muller et al. (2011) The discovery of 5-formylcytosine in embryonic stem cell DNA. *Angew Chem Int Ed Engl* **50**:7008–7012.
29. L Zhang, X Lu, J Lu, H Liang, Q Dai et al. (2012) Thymine DNA glycosylase specifically recognizes 5-carboxylcytosine-modified DNA. *Nat Chem Biol* **8**:328–330.
30. JU Guo, Y Su, C Zhong, GL Ming, H Song. (2011) Hydroxylation of 5-methylcytosine by TET1 promotes active DNA demethylation in the adult brain. *Cell* **145**:423–434.
31. A Szwagierczak, S Bultmann, CS Schmidt, F Spada, H Leonhardt. (2010) Sensitive enzymatic quantification of 5-hydroxymethylcytosine in genomic DNA. *Nucleic Acids Res* **38**:e181.

32. S Kriaucionis, N Heintz. (2009) The nuclear DNA base 5-hydroxymethylcytosine is present in Purkinje neurons and the brain. *Science* **324**:929–930.
33. Y Xu, F Wu, L Tan, L Kong, L Xiong *et al.* (2011) Genome-wide regulation of 5hmC, 5mC, and gene expression by TET1 hydroxylase in mouse embryonic stem cells. *Mol Cell* **42**:451–464.
34. H Wu, AC D'Alessio, S Ito, Z Wang, K Cui *et al.* (2011) Genome-wide analysis of 5-hydroxymethylcytosine distribution reveals its dual function in transcriptional regulation in mouse embryonic stem cells. *Genes Dev* **25**:679–684.
35. WA Pastor, UJ Pape, Y Huang, HR Henderson, R Lister *et al.* (2011) Genome-wide mapping of 5-hydroxymethylcytosine in embryonic stem cells. *Nature* **473**:394–397.
36. G Ficz, MR Branco, S Seisenberger, F Santos, F Krueger *et al.* (2011) Dynamic regulation of 5-hydroxymethylcytosine in mouse ES cells and during differentiation. *Nature* **473**:398–402.
37. V Valinluck, LC Sowers. (2007) Endogenous cytosine damage products alter the site selectivity of human DNA maintenance methyltransferase DNMT1. *Cancer Res* **67**:946–950.
38. H Wu, AC D'Alessio, S Ito, K Xia, Z Wang *et al.* (2011) Dual functions of TET1 in transcriptional regulation in mouse embryonic stem cells. *Nature* **473**:389–393.
39. K Williams, J Christensen, MT Pedersen, JV Johansen, PA Cloos *et al.* (2011) TET1 and hydroxymethylcytosine in transcription and DNA methylation fidelity. *Nature* **473**:343–348.
40. W Reik, W Dean, J Walter. (2001) Epigenetic reprogramming in mammalian development. *Science* **293**:1089–1093.
41. ZD Smith, MM Chan, TS Mikkelsen, H Gu, A Gnirke *et al.* (2012) A unique regulatory phase of DNA methylation in the early mammalian embryo. *Nature* **484**:339–344.
42. J Borgel, S Guibert, Y Li, H Chiba, D Schubeler *et al.* (2010) Targets and dynamics of promoter DNA methylation during early mouse development. *Nat Genet* **42**:1093–1100.
43. SA Smallwood, S Tomizawa, F Krueger, N Ruf, N Carli *et al.* (2011) Dynamic CpG island methylation landscape in oocytes and preimplantation embryos. *Nat Genet* **43**:811–814.

44. N Lane, W Dean, S Erhardt, P Hajkova, A Surani et al. (2003) Resistance of IAPs to methylation reprogramming may provide a mechanism for epigenetic inheritance in the mouse. *Genesis* **35**:88–93.
45. CR Farthing, G Ficz, RK Ng, CF Chan, S Andrews et al. (2008) Global mapping of DNA methylation in mouse promoters reveals epigenetic reprogramming of pluripotency genes. *PLoS Genet* **4**:e1000116.
46. P Hajkova, K Ancelin, T Waldmann, N Lacoste, UC Lange et al. (2008) Chromatin dynamics during epigenetic reprogramming in the mouse germ line. *Nature* **452**:877–881.
47. J Bednar, RA Horowitz, SA Grigoryev, LM Carruthers, JC Hansen et al. (1998) Nucleosomes, linker DNA, and linker histone form a unique structural motif that directs the higher-order folding and compaction of chromatin. *Proc Natl Acad Sci USA* **95**:14173–14178.
48. DE Olins, AL Olins. (2003) Chromatin history: our view from the bridge. *Nat Rev Mol Cell Biol* **4**:809–814.
49. AJ Bannister, T Kouzarides. (2011) Regulation of chromatin by histone modifications. *Cell Res* **21**:381–395.
50. P Tessarz, T Kouzarides. (2014) Histone core modifications regulating nucleosome structure and dynamics. *Nat Rev Mol Cell Biol* **15**:703–708.
51. BM Turner. (2000) Histone acetylation and an epigenetic code. *Bioessays* **22**:836–845.
52. VG Allfrey, R Faulkner, AE Mirsky. (1964) Acetylation and methylation of histones and their possible role in the regulation of RNA synthesis. *Proc Natl Acad Sci USA* **51**:786–794.
53. M Oki, H Aihara, T Ito. (2007) Role of histone phosphorylation in chromatin dynamics and its implications in diseases. *Subcell Biochem* **41**:319–336.
54. MT Bedford. (2007) Arginine methylation at a glance. *J Cell Sci* **120**:4243–4246.
55. F Fuks, PJ Hurd, R Deplus, T Kouzarides. (2003) The DNA methyltransferases associate with HP1 and the SUV39H1 histone methyltransferase. *Nucleic Acids Res* **31**:2305–2312.
56. JC Rice, SD Briggs, B Ueberheide, CM Barber, J Shabanowitz et al. (2003) Histone methyltransferases direct different degrees of methylation to define distinct chromatin domains. *Mol Cell* **12**:1591–1598.

57. SS Levine, A Weiss, H Erdjument-Bromage, Z Shao, P Tempst *et al.* (2002) The core of the polycomb repressive complex is compositionally and functionally conserved in flies and humans. *Mol Cell Biol* **22**:6070–6078.
58. AP Bracken, K Helin. (2009) Polycomb group proteins: navigators of lineage pathways led astray in cancer. *Nat Rev Cancer* **9**:773–784.
59. L Di Croce, K Helin. (2013) Transcriptional regulation by polycomb group proteins. *Nat Struct Mol Biol* **20**:1147–1155.
60. BE Bernstein, TS Mikkelsen, X Xie, M Kamal, DJ Huebert *et al.* (2006) A bivalent chromatin structure marks key developmental genes in embryonic stem cells. *Cell* **125**:315–326.
61. P Voigt, WW Tee, D Reinberg. (2013) A double take on bivalent promoters. *Genes Dev* **27**:1318–1338.
62. H Richly, L Aloia, L Di Croce. (2011) Roles of the polycomb group proteins in stem cells and cancer. *Cell Death Dis* **2**:e204.
63. JE Ohm, KM McGarvey, X Yu, L Cheng, KE Schuebel *et al.* (2007) A stem cell-like chromatin pattern may predispose tumor suppressor genes to DNA hypermethylation and heritable silencing. *Nat Genet* **39**:237–242.
64. Y Shi, F Lan, C Matson, P Mulligan, JR Whetstine *et al.* (2004) Histone demethylation mediated by the nuclear amine oxidase homolog LSD1. *Cell* **119**:941–953.
65. JR Whetstine, A Nottke, F Lan, M Huarte, S Smolikov *et al.* (2006) Reversal of histone lysine trimethylation by the JMJD2 family of histone demethylases. *Cell* **125**:467–481.
66. K Sakabe, GW Hart. (2010) O-GlcNAc transferase regulates mitotic chromatin dynamics. *J Biol Chem* **285**:34460–34468.
67. PO Hassa, SS Haenni, M Elser, MO Hottiger. (2006) Nuclear ADP-ribosylation reactions in mammalian cells: where are we today and where are we going? *Microbiol Mol Biol Rev* **70**:789–829.
68. R Martinez-Zamudio, HC Ha. (2012) Histone ADP-ribosylation facilitates gene transcription by directly remodeling nucleosomes. *Mol Cell Biol* **32**:2490–2502.
69. S Messner, M Altmeyer, H Zhao, A Pozivil, B Roschitzki *et al.* (2010) PARP1 ADP-ribosylates lysine residues of the core histone tails. *Nucleic Acids Res* **38**:6350–6362.

70. T Kouzarides. (2007) Chromatin modifications and their function. *Cell* **128**:693–705.
71. M Lachner, D O'Carroll, S Rea, K Mechtler, T Jenuwein. (2001) Methylation of histone H3 lysine 9 creates a binding site for HP1 proteins. *Nature* **410**:116–120.
72. AJ Bannister, P Zegerman, JF Partridge, EA Miska, JO Thomas et al. (2001) Selective recognition of methylated lysine 9 on histone H3 by the HP1 chromo domain. *Nature* **410**:120–124.
73. RA Gupta, N Shah, KC Wang, J Kim, HM Horlings et al. (2010) Long non-coding RNA HOTAIR reprograms chromatin state to promote cancer metastasis. *Nature* **464**:1071–1076.
74. T Hung, HY Chang. (2010) Long noncoding RNA in genome regulation: prospects and mechanisms. *RNA Biol* **7**:582–585.
75. W Reik, J Walter. (2001) Genomic imprinting: parental influence on the genome. *Nat Rev Genet* **2**:21–32.
76. CP Walsh, JR Chaillet, TH Bestor. (1998) Transcription of IAP endogenous retroviruses is constrained by cytosine methylation. *Nat Genet* **20**:116–117.
77. RA Waterland, RL Jirtle. (2003) Transposable elements: targets for early nutritional effects on epigenetic gene regulation. *Mol Cell Biol* **23**:5293–5300.
78. MS Bartolomei, AC Ferguson-Smith. (2011) Mammalian genomic imprinting. *Cold Spring Harb Perspect Biol* **3**.
79. MG Butler. (2009) Genomic imprinting disorders in humans: a minireview. *J Assist Reprod Genet* **26**:477–486.
80. DB Pontier, J Gribnau. (2011) Xist regulation and function explored. *Hum Genet* **130**:223–236.
81. D Tian, S Sun, JT Lee. (2010) The long noncoding RNA, Jpx, is a molecular switch for X chromosome inactivation. *Cell* **143**:390–403.
82. KM Godfrey, KA Lillycrop, GC Burdge, PD Gluckman, MA Hanson. (2007) Epigenetic mechanisms and the mismatch concept of the developmental origins of health and disease. *Pediatr Res* **61**: 5R–10R.
83. C Gemma, S Sookoian, J Alvarinas, SI Garcia, L Quintana et al. (2009) Maternal pregestational BMI is associated with methylation of the PPARGC1A Promoter in newborns. *Obesity* (SilverSpring).

84. MK Skinner. (2014) Endocrine disruptor induction of epigenetic transgenerational inheritance of disease. *Mol Cell Endocrinol* **398**:4–12.
85. M Szyf, P McGowan, MJ Meaney. (2008) The social environment and the epigenome. *Environ Mol Mutagen* **49**:46–60.
86. PD Gluckman, MA Hanson, HG Spencer, P Bateson. (2005) Environmental influences during development and their later consequences for health and disease: implications for the interpretation of empirical studies. *Proc Biol Sci* **272**:671–677.
87. SA Kliewer, BM Forman, B Blumberg, ES Ong, U Borgmeyer et al. (1994) Differential expression and activation of a family of murine peroxisome proliferator-activated receptors. *Proc Natl Acad Sci USA* **91**:7355–7359.
88. C Canto, J Auwerx. (2009) PGC-1alpha, SIRT1 and AMPK, an energy sensing network that controls energy expenditure. *Curr Opin Lipidol* **20**:98–105.
89. R Feil, MF Fraga. (2011) Epigenetics and the environment: emerging patterns and implications. *Nat Rev Genet* **13**:97–109.
90. YI Kim. (2005) Nutritional epigenetics: impact of folate deficiency on DNA methylation and colon cancer susceptibility. *J Nutr* **135**:2703–2709.
91. LA Racey, P Byvoet. (1971) Histone acetyltransferase in chromatin. Evidence for in vitro enzymatic transfer of acetate from acetyl-coenzyme A to histones. *Exp Cell Res* **64**:366–370.
92. L Cai, BM Sutter, B Li, BP Tu. (2011) Acetyl-CoA induces cell growth and proliferation by promoting the acetylation of histones at growth genes. *Mol Cell* **42**:426–437.
93. F Forneris, C Binda, MA Vanoni, A Mattevi, E Battaglioli. (2005) Histone demethylation catalysed by LSD1 is a flavin-dependent oxidative process. *FEBS Lett* **579**:2203–2207.
94. JL Meier. (2013) Metabolic mechanisms of epigenetic regulation. *ACS Chem Biol* **8**:2607–2621.

Nutrition, Epigenetics and the Early Life Origins of Disease: Evidence from Human Studies

CHAPTER 2

Nina D'Vaz* and Rae-Chi Huang[†]

*School of Paediatrics and Child Health (M561),
[†]Telethon Kids Institute (M560),
The University of Western Australia,
35 Stirling Highway,
Crawley, WA 6009, Australia

Introduction

Early life origins of disease are predicated on the observations that *in utero* and potentially periconceptional environment is associated with the later development of offspring non-communicable diseases (NCDs) including obesity, coronary artery disease, diabetes, mental health, cancer and allergic disorders. While the mechanisms of environmental influence are still somewhat unclear, evidence is increasingly in support of environmental factors influencing offspring phenotype (at least partially) through modification of the developing fetal epigenome. As reviewed by Lillycrop and Burdge,[1] epigenetic modifications have been shown to occur throughout life from conception through to adulthood, but the fetal epigenome is showing the greatest plasticity and sensitivity to modification during the establishment of methylation patterns.

Environmental factors influencing the *in utero* and periconceptional milieu are clearly multifactorial and somewhat interdependent and it is difficult to separate the effect of one from those of others. However, one factor, which is relatively easily controlled and has been studied in animal and human studies is maternal nutrition. As discussed in Chapter 3, animal studies have established a causative link between maternal nutrition, offspring epigenetic changes and offspring phenotype and have formed the basis for investigating the same associations in humans, through observational and intervention based studies worldwide.

Broadly speaking there is increasing evidence of the role of epigenetics in early life origins of disease in humans. While this does not specifically address whether nutrition is the cause of or contributor to changes in epigenetics and phenotype, several observations lend credence to this hypothesis and a new rapidly expanding research field "nutritional genomics" is committed to unravelling the interaction between nutrition, genes and epigenetics.

Maternal Nutrition

Evidence for the Effect of Maternal Nutrition on Epigenetic Processes in the Early Life Origins of Disease

Famine and maternal under-nutrition

The evidence for maternal nutrition inducing offspring epigenetic changes and disease is mostly indirect in humans and has mainly come from "naturally" occurring events rather than designed interventions.

The best known evidence is from the Dutch Hunger Winter which affected part of the Netherlands from November 1944 to June 1945. During the 2nd World War, Germany occupied parts of the Netherlands and food supply was restricted to daily rations as low as 400 calories per day. As reviewed by Painter *et al.*,[2] this unfortunate occurrence of extreme food shortage provided ta unique opportunity to study the effect of sudden extreme maternal under-nutrition during pregnancy on offspring outcomes in an otherwise "healthy" population. When compared to "normal" children not affected by "starvation in the womb" (born before or more than nine months after the Hunger Winter), offspring affected by the hunger period during the antenatal period, showed several abnormalities in epigenotypes and phenotypic outcomes.

Altered DNA methylation leading to fetal programming was first observed at the IGF2/H19 locus, which encodes Insulin-like growth factor 2. IGF2 highly influences human growth and development. Offspring who were exposed to maternal famine during early gestation had decreased methylation in a region controlling expression within the IGF2 locus accompanied by increased obesity, 60 years after the exposure. Interestingly this was not seen if offspring had been affected during mid-late gestation. This indicates that gestational timing, potentially reflecting the timing of development and maturation of fetal organ systems, is indeed a factor in determining the effect of environmental exposures.

Later, differential methylation of a further six genes associated with growth and metabolic/cardiovascular phenotypes (IL10, GNASAS, INSIGF, LEP, ABCA1 and MEG3) was shown in offspring to mothers who experienced the Dutch famine during periconception.[3] Although these observational findings require further detailed and well-controlled investigations, it is clear that the human epigenome is susceptible to environmental modulation during early gestation. It is thought that this period of early fetal development offers a unique window of epigenetic plasticity that may be utilized for nutritional interventions aimed at optimizing offspring phenotypes and well-designed human studies are called for.

Other indirect evidence has emerged recently from studies in the Gambia which have shown variation in infant methylation status according to the season of pregnancy. In rural Gambia variation in diet accompany the seasonal climate variations, monomodal "hungry" rainy season and dry harvest season. The population's primary dependence on the consumption of own-grown foods, leads to profound annual variations in the intakes of macro and micronutrients. Somewhat counter to intuition, Dominguez-Salas et al.[4] observed elevated DNA methylation at presumed metastable epialleles in children conceived in the protein-energy-limited rainy ("hungry") season compared to those conceived in the dry ("harvest") season. The authors hypothesized and have in turn shown that, rather than energy intake per se, the specific availability of one-carbon nutrients critical to methyl-donor metabolic pathways, which somewhat paradoxically increases in the rainy season, account for the observed variability in methylation.[4]

Contrary to the Dutch famine, conception during the Gambian hungry season did not associate with altered methylation of the IGF2 gene, which may reflect the difference in severity and environmental context of these two famine scenarios.

Some evidence of effect of moderate under-nutrition on fetal programming does exist from interventional studies in humans. Offspring born to mothers who experienced significant weight loss as a result of biliopancreatic diversion bariatric surgery showed improvements in cardiometabolic markers sustained into adolescence.[5] This effect was attributed to an improved intrauterine environment.[6] It is still unclear if epigenetic changes underpin the observed phenotypic effects in the offspring. Unravelling the molecular mechanism for this change in offspring is a research priority.

Obesity in the mother

While the Dutch and Gambian studies demonstrate interesting effects of under-nutrition on offspring epigenetic profiles and phenotypes, it is perhaps the opposite scenario of over-nutrition and obesity that is more relevant to the human population. Both developed and third world countries face an escalating obesity crisis. Inducing obesity or over-nutrition however, is unethical and impractical. While over-nutrition scenarios have been investigated in animal models (discussed elsewhere), studies of the effect of human over-nutrition on offspring epigenotype are negligible. Several observations however do point toward an effect of maternal obesity on childhood health outcomes.

Maternal obesity, significant weight gain or high dietary fat and sugar intake during pregnancy is associated with pathologically altered maternal factors including glucose metabolism and this may induce changes in fetal growth and developmental trajectories.[6]

As such, human studies have shown that maternal obesity may predispose offspring to develop obesity, diabetes, cardiovascular abnormalities, adult schizophrenia and asthma.[5,7–12] The mechanisms which have been proposed for these effects are hyperglycemia, arachidonic acid-prostaglandin and nitric oxide pathways, increased reactive oxygen species, lipotoxicity and placental function, hyperinsulinemia and insulin resistance.

While studies do show an effect of maternal obesity on childhood outcomes, it remains speculative whether this effect is mediated by epigenetic mechanisms and well-designed human studies are called for.

Specific Dietary Components

While many studies have focused on overall caloric load as the determining factor of epigenetic changes, the Gambian study[4] highlights the possibility that rather than absolute energy intake, availability of specific dietary components may be of importance.

Broadly speaking, specific nutritional components of interest to epigenetic manipulation might be categorized into methyl donors, fatty acids, vitamins, prebiotics, amongst other miscellaneous components.

Methyl Donors and Cofactors in 1-carbon Metabolism

Of specific interest are co-actors such as vitamin B12 and folate, and methyl donors including choline, betaine, methionine, serine, glycine and histidine. Of these, folate has received the most attention as a potent methyl donor for one-carbon metabolism, which assists DNA methylation, and the level of bioavailable folate may support or restrict methylation activity.

Folate and Folic Acid

As discussed in Chapter 3, animal studies have very convincingly shown epigenetic effects of dietary folic acid intake on phenotypic traits. The relatively recent introduction of folate supplement recommendations during pregnancy to prevent neural defects has provided opportunity to study epigenetic effects of folic acid intake in some detail.

In 2009, Steegers-Theunissen et al.[13] showed that maternal folic acid intake was associated with 4.5% increase in IGF2 differentially methylated region (DMR) methylation in the child. The same study also showed that IGF2 DMR methylation of the children was associated with the concentration of S-adenosylmethionine in the blood of the mother, but not of the child (+1.7% methylation per SD S-adenosylmethionine; p = 0.037). More recently, in 463 mother-infant pairs from the large-scale GenR study, van Mill et al.[14] showed that maternal folate deficiency is associated with reduced infant methylation status, but it is unclear whether this had any effect on infant health outcomes. Similarly, Hoyo et al.[15] detected an inverse relationship between maternal folic acid intake before and during pregnancy and H19 DNA methylation. A further smaller study compared women with high (n = 11) folate to those with low (n = 12) folate levels during the last trimester of pregnancy and found hypomethyaltion of a 923bp region 3kb upstream of at a region of ZFP57 region, a regulator of DNA methylation during development. In CD4(+) cells, hypomethylation was associated with increased H3/H4 acetylation at the ZFP57 promoter and higher mRNA expression.[16]

However, the association of DNA methylation with maternal folate has not been consistently observed. Some follow up studies have shown no association between folate status and DNA methylation (either globally or site-specific).[17–19]

Choline

Choline is a methyl group donor which is synthesized endogenously, but also obtained from the diet. Many of the foods rich in choline are also high in fats; therefore, decreasing intakes have been seen so that many do not now achieve recommended intake levels.[20]

A small study showed that maternal supply of choline has been associated with differences in DNA methylation of cortisol-regulating genes in placenta and cord leukocytes.

Higher maternal choline intake yielded higher placental promoter methylation of the cortisol-regulating genes, corticotropin releasing hormone (*CRH*; $P = 0.05$) and glucocorticoid receptor (*NR3C1*; $P = 0.002$); lower placental *CRH* transcript abundance ($P = 0.04$); lower cord blood leukocyte promoter methylation of *CRH* ($P = 0.05$) and *NR3C1* ($P = 0.04$); and 33% lower ($P = 0.07$) cord plasma cortisol. In addition, placental global DNA methylation and dimethylated histone H3 at lysine 9 (H3K9me2) were higher ($P = 0.02$) in the high choline group.[21] These early data will need further confirmation in larger numbers of human subjects.

Dietary Fat

The effect of dietary fat on epigenetic outcomes will be discussed in Chapter 4.

Vitamin D

Vitamin D levels are highly linked to sun exposure and skin pigmentation, and a modern tendency for more sedentary and indoor behaviors is causing an increasing incidence of vitamin D deficiencies. Vitamin D is supplied in relatively small amounts through the diet and foods rich in dietary vitamin D include fatty fish and cod liver oil. With modern diets including less oily fish, foods fortified with vitamin D (including milk and margarine) as well as the use of oral vitamin D supplementation is now common in some populations.[22] A possible link between maternal vitamin D deficiency to increasing NCD occurrence has been suggested, possibly mediated through epigenetic mechanisms.

A recent study showed that hypermethylation of key placental genes involved in vitamin D metabolism was associated with preeclampsia and reduced expression of Retinoid X receptor α (RXR).[23] Interestingly, the methylation of a CpG site in the promoter region of the nuclear receptor, RXRA in umbilical cord DNA was strongly related to childhood adiposity in both boys and girls in two independent cohorts.[24]

Largely however, the evidence for a role of vitamin D in epigenetically regulating phenotypes is not established and currently remains speculative, requiring further investigations before confirmation of this link.

Prebiotics, Short Chain Fatty Acids and Gut Microbiota

Along with the changes mentioned above, modern diets also contain less fibre and increased processed foods than historically. This has led to changes in gut microbiota biodiversity which has been linked to obesity[25] and there is now emerging interest in the role of the microbiome on altering epigenetic profiles.

In a study by Kumar and colleagues,[26] gut microbiota profiles, with either Firmicutes (n = 4) or Bacteroides (n = 4) as a dominant group in pregnant women, showed a correlation with differential methylation status of gene promoters functionally associated with cardiovascular diseases from whole blood DNA collected six months post-partum.

Once again, these associations need to be replicated in larger numbers before confirmation of the association of the gut microbiota with epigenetic modification.

Dietary Patterns

In conclusion, although studies show some associations between specific maternal nutrient intake and DNA methylation, variation in study design as well as the likely interactive relationships between nutrients cloud the picture of possible causal relationships, but there is certainly evidence to suggest some effect of maternal dietary components on the epigenome of both mothers and offspring.

Given the likely interaction of individual nutrients, it makes practical sense to identify epigenetic effects of dietary patterns rather than isolated dietary components. In support of this approach, indirect evidence in humans includes the observation that a "healthy" dietary pattern protects against the global DNA methylation that is associated with an unhealthy "Western" dietary pattern.[27] Mediterranean diets are high in monounsaturated fatty acids, antioxidant vitamins and phytochemicals.[28] The protective effect of this "healthy" diet for cardiovascular disease and cancer may be mediated by epigenetic mechanisms.[29]

Paternal Nutrition

Evidence for the Role of Paternal Nutrition on Epigenetics

Adding further complexity, it is becoming increasingly thought that paternal "nutritional status" also interacts with offspring epigenetic marks. This was initially shown in a rat model which is outlined in detail elsewhere.[30] This, in brief, showed that a high fat diet in the father could directly alter the offspring epigenetics by paternal transmission and evidence is emerging to indicate that this phenomenon may occur in humans too. In 2013 Soubry et al.[31] showed that hypomethylation of IGF2 DMR in cord blood leukocyte DNA from 79 newborns was associated with obesity in the father. This effect remained significant after adjustment for maternal and newborn characteristics. More recently these findings have been followed up by the same group with a study, which show further epigenetic loci including mesoderm-specific transcript (MEST) and neuronatin (NNAT) were also hypomethylated in those whose fathers were obese.[32]

The potential mechanisms for origin of male obesity-induced paternal programming include: (1) Accumulation of sperm DNA damage resulting in de novo mutations in the embryo; (2) changes in the sperm epigenetic marks (microRNA, methylation or acetylation).

So human association studies are suggestive of a paternal epigenomic inheritance but have yet to show conclusively that paternal nutrition per se (be it high fat diet, or excess calories) mediates newborn methylation changes.

Epigenetic Biomarkers of Disease Risk in Humans

It has been observed by Zhu et al.[33] that DNA methylation patterns vary with adipocyte differentiation, suggesting epigenetics play a role in adipogenesis and several studies have shown associations between DNA methylation in neonatal tissues and childhood adiposity.[24,34] Godfrey and colleagues[24] further found that the methylation of a CpG site in the promoter region of the nuclear receptor, Retinoid X receptor α (RXRA) in umbilical cord DNA was strongly related to childhood adiposity in both boys and girls in two independent cohorts and RXRA promoter methylation explained significant variance (>25%) in childhood fat mass. Linking this to maternal diet in pregnancy, there was an association with higher methylation of RXRA lower maternal carbohydrate intake in early pregnancy, previously linked with higher neonatal adiposity in this population. In 204 infants it was further observed by Perkins and colleagues[34] that levels of H19 methylation (measured by pyrosequencing at birth) were higher in those who were heavier at 12 months of age. Those with weight for age >85th percentile had on average 3.4% higher DNA methylation at the H19 loci.

While these studies show a predictive effect of birth epigenetics on childhood obesity, these associations continue to later life when childhood methylation patterns correlate with obesity outcomes. As such, the methylation of specific CpG loci in the promoter of PGC1α, at five years of age predict adiposity year-on-year from eight to fourteen years[35] and IGF2/H19 methylation is associated with subcutaneous fat thickness in 17 year olds.[36] The associations between DNA methylation and obesity increase to adulthood and as an example, in Korean women, interleukin 6 DNA methylation has been shown to be elevated in obese, compared with normal weight or overweight women.[37]

These associations between DNA methylation and adiposity have been demonstrated by measuring DNA methylation in peripheral tissues (whole blood,[35,36] leukocytes, cord blood[34] and umbilical cord[24]). Use of peripheral tissue is an issue that needs consideration in general in human studies as DNA methylation patterns vary between tissues, Peripheral tissue may not accurately reflect the status of the tissue of actual interest. However, as it is often impractical and unethical to collect the target tissue of interest (adipose, pancreatic, liver and muscle tissue), establishing the stability of epigenetic marks between different tissues is urgently needed.

Knowledge Gaps and Future Directions

The evidence for the role of nutrition on epigenetics in humans is indirect, but supportive of animal studies. While suggestive, more direct evidence will be needed to ascertain firstly that altering maternal diet can change offspring epigenetics. This might be achieved in the future through randomized controlled trials which alter maternal (and paternal) diet, during or prior to pregnancy. Second, human studies need to be able to show that altering offspring epigenetics also results in changes in offspring phenotype and downstream functional modifications. Some of this may be achievable with *in vivo* experiments, however, much definitive evidence will be some time in coming via randomized controlled trials.

Currently the field in human studies is greatly focused on DNA methylation. There are many other types of epigenetics. These other forms, as discussed above may have just as large effect sizes on function, which may become more apparent as they are increasingly explored.

Tissue availability limits interpretation of most human studies. Most commonly blood cells are used for methylation assessment, but it is not known how well this tissue represents other, perhaps more relevant, body compartments. In the case of obesity, experiments in adipose, pancreatic, muscle and liver tissue would more accurately reflect the relevant functional changes, than whole blood DNA. It should be taken into consideration however, that to be of any practical use for screening purposes, epigenetic changes must be detectable in readily available tissues (like whole blood or buccal swabs) as screening will otherwise not be feasible on a population scale. Therefore, as discussed earlier in this chapter, a comprehensive library documenting the stability of epigenetic marks between different tissues is urgently needed. Concurrently, this field of research in humans needs to refine the corrections that can be undertaken to adjust for heterogeneity of cell type in tissues of convenience (such as whole blood and buccal cells) at the time of analysis.[38,39]

A further development needed to support this field of research is a more comprehensive understanding of the manner in which

DNA methylation may fluctuate or change over a person's lifetime. It is likely that some sites are stable, where others change, either in a gradual decline with age or fluctuate with other influences. Some studies indicate that even brief environmental changes including acute exercise[40] can change methylation profiles in muscle tissues and such sensitivity to change may make it difficult to decipher effects of individual interventions.

Hitherto, much research has emphasized the direct role of maternal nutrition. The role of the father's nutrition is also a promising future research direction that will need to be understood to fill the gaps in knowledge in trans-generational effects of epigenetics. To understand this further, research on sperm, and epidemiology studies of multiple generations will be invaluable to understand intergenerational effects through the paternal line.[41]

Understanding epigenetics may have several potential future benefits. If epigenetic mechanisms are indeed causally associated with NCDs and sensitive to environmental modulation, the possibility of therapeutic epigenome modulation for disease prevention or treatment exists.

Indeed, inhibitors of DNA methylation and acetylase inhibitors have been approved for cancer treatment as they improve cell survival and show less toxicity than conventional chemotherapy drugs. However, the lack of tissue and cell line specificity limit the therapeutic scope of these compounds in disorders of less acute severity. In conditions like obesity, the use of less drastic functional foods as epigenome modifiers may be more appropriate.

A more realistic direction that this field will progress in is to contribute to personalized or precision medicine for the management of obesity and related diseases. This is likely to be an incremental process whereby we develop increasing understanding of how epigenetic markers relate to nutrition and interact with other factors within the individual such as (genetic makeup and sex). An early rudimentary example of this the demonstration of a sex difference in how dietary omega-3 fatty acid supplementation affects DNA methylation.[42]

Conclusion

The epigenome is increasingly thought to provide a dynamic interface between the ever changing environment and the genome. While the epigenome displays plasticity throughout the life course, it is thought that the time of greatest plasticity is during early fetal development when methylation patterns are undergoing establishment. If indeed epigenotypes are modifiable by pregnancy milieu and epigenetic variance affects NCD predisposition, this places extreme importance on pregnancy diet and behavior of both mothers and fathers to ensure the best possible epigenetic makeup of their children. It is tempting to speculate that in a true sense, that not only are you what you eat, your children are too.

References

1. KA Lillycrop, GC Burdge. (2015) Maternal diet as a modifier of offspring epigenetics. *J Dev Orig Health Dis* **6**(2):88–95.
2. RC Painter, TJ Roseboom, OP Bleker. (2015) Prenatal exposure to the Dutch famine and disease in later life: an overview. *Reprod Toxicol* **20**(3):345–352.
3. BT Heijmans *et al.* (2009) The epigenome: archive of the prenatal environment. *Epigenetics* **4**(8):26–531.
4. P Dominguez-Salas *et al.* (2014) Maternal nutrition at conception modulates DNA methylation of human metastable epialleles. *Nat Commun* **5**:3746.
5. J Smith *et al.* (2009) Effects of maternal surgical weight loss in mothers on intergenerational transmission of obesity. *J Clin Endocrinol Metab* **94**(11):4275–4283.
6. D Dowling, FM McAuliffe. (2013) The molecular mechanisms of offspring effects from obese pregnancy. *Obes Facts* **6**(2):134–145.
7. DJ Barker. (2007) The origins of the developmental origins theory. *J Intern Med* **261**(5):412–417.
8. MF Higgins *et al.* (2008) Fetal anterior abdominal wall thickness in diabetic pregnancy. *Eur J Obstet Gynecol Reprod Biol* **140**(1):43–47.

9. SJ Herring, E Oken. (2011) Obesity and diabetes in mothers and their children: can we stop the intergenerational cycle? *Curr Diab Rep* **11**(1):20–27.
10. GM Khandaker, CR Dibben, PB Jones. (2012) Does maternal body mass index during pregnancy influence risk of schizophrenia in the adult offspring? *Obes Rev* **13**(6):518–527.
11. SP Patel *et al.* (2012) Associations between pre-pregnancy obesity and asthma symptoms in adolescents. *J Epidemiol Community Health* **66**(9):809–814.
12. DJ Barker. (2007) Obesity and early life. *Obes Rev* **8 Suppl 1**(s1): 45–49.
13. RP Steegers-Theunissen *et al.* (2009) Periconceptional maternal folic acid use of 400 microg per day is related to increased methylation of the IGF2 gene in the very young child. *PLoS One* **4**(11):e7845.
14. NH van Mil *et al.* (2014) Determinants of maternal pregnancy one-carbon metabolism and newborn human DNA methylation profiles. *Reproduction* **148**(6):581–592.
15. C Hoyo *et al.* (2011) Methylation variation at IGF2 differentially methylated regions and maternal folic acid use before and during pregnancy. *Epigenetics* **6**(7):928–936.
16. M Amarasekera *et al.* (2014) Genome-wide DNA methylation profiling identifies a folate-sensitive region of differential methylation upstream of ZFP57-imprinting regulator in humans. *FASEB J* **28**(9):4068–4076.
17. AA Fryer *et al.* (2009) LINE-1 DNA methylation is inversely correlated with cord plasma homocysteine in man: a preliminary study. *Epigenetics* **4**(6):394–398.
18. AA Fryer *et al.* (2011) Quantitative, high-resolution epigenetic profiling of CpG loci identifies associations with cord blood plasma homocysteine and birth weight in humans. *Epigenetics* **6**(1):86–94.
19. Y Ba *et al.* (2011) Relationship of folate, vitamin B12 and methylation of insulin-like growth factor-II in maternal and cord blood. *Eur J Clin Nutr* **65**(4):480–485.
20. HH Jensen *et al.* (2007) Choline in the diets of the US population: NHANES, 2003–2004. *The FASEB Journal* **21**(6):LB46.
21. X Jiang *et al.* (2012) Maternal choline intake alters the epigenetic state of fetal cortisol-regulating genes in humans. *FASEB J* **26**(8): 3563–3574.

22. A Hossein-nezhad, MF Holick. (2012) Optimize dietary intake of vitamin D: an epigenetic perspective. *Curr Opin Clin Nutr Metab Care* **15**(6):567–579.
23. CM Anderson *et al.* (2015) First trimester vitamin D status and placental epigenomics in preeclampsia among Northern Plains primiparas. *Life Sci* **129**:10–15.
24. KM Godfrey *et al.* (2011) *Epigenetic gene promoter methylation at birth is associated with child's later adiposity. Diabetes* **60**(5):1528–1534.
25. PJ Turnbaugh *et al.* (2006) An obesity-associated gut microbiome with increased capacity for energy harvest. *Nature* **444**(7122):1027–1031.
26. H Kumar *et al.* (2014) Gut microbiota as an epigenetic regulator: pilot study based on whole-genome methylation analysis. *MBio* **5**(6).
27. FF Zhang *et al.* (2011) Dietary patterns are associated with levels of global genomic DNA methylation in a cancer-free population. *J Nutr* **141**(6):1165–1171.
28. D Corella, JM Ordovas. (2014) How does the Mediterranean diet promote cardiovascular health? Current progress toward molecular mechanisms: gene-diet interactions at the genomic, transcriptomic, and epigenomic levels provide novel insights into new mechanisms. *Bioessays* **36**(5):526–537.
29. FI Milagro *et al.* (2013) Dietary factors, epigenetic modifications and obesity outcomes: progresses and perspectives. *Mol Aspects Med* **34**(4):782–812.
30. SF Ng *et al.* (2010) Chronic high-fat diet in fathers programs beta-cell dysfunction in female rat offspring. *Nature* **467**(7318):963–966.
31. A Soubry *et al.* (2013) Paternal obesity is associated with IGF2 hypomethylation in newborns: results from a Newborn Epigenetics Study (NEST) cohort. *BMC Med* **11**:29.
32. A Soubry *et al.* (2015) Newborns of obese parents have altered DNA methylation patterns at imprinted genes. *Int J Obes (Lond)* **39**(4): 650–657.
33. JG Zhu *et al.* (2012) Differential DNA methylation status between human preadipocytes and mature adipocytes. *Cell Biochem Biophys* **63**(1):1–5.
34. E Perkins *et al.* (2012) Insulin-like growth factor 2/H19 methylation at birth and risk of overweight and obesity in children. *J Pediatr* **161**(1):31–39.

35. R Clarke-Harris, WT Hosking, J Pinkney, AN Jeffery, BS Metcalf, KM Godfrey, LD Voss, KA Lillycrop, GC Burdge. (2014) Peroxisomal proliferator activated receptor-γ-co-activator-1α promoter methylation in blood at 5–7 years predicts adiposity from 9 to 14 years (EarlyBird 50). *Diabetes*.
36. RC Huang *et al.* (2012) DNA methylation of the IGF2/H19 imprinting control region and adiposity distribution in young adults. *Clin Epigenetics* **4**(1):21.
37. YK Na *et al.* (2015) Increased methylation of interleukin 6 gene is associated with obesity in Korean women. *Mol Cells* **38**(5):452–456.
38. BT Adalsteinsson *et al.* (2012) Heterogeneity in white blood cells has potential to confound DNA methylation measurements. *PLoS One* **7**(10):e46705.
39. EA Houseman *et al.* (2008) Model-based clustering of DNA methylation array data: a recursive-partitioning algorithm for high-dimensional data arising as a mixture of beta distributions. *BMC Bioinformatics* **9**:365.
40. R Barres *et al.* (2012) Acute exercise remodels promoter methylation in human skeletal muscle. *Cell Metab* **15**(3):405–411.
41. A Soubry *et al.* (2014) A paternal environmental legacy: evidence for epigenetic inheritance through the male germ line. *Bioessays* **36**(4):359–371.
42. SP Hoile *et al.* (2014) Supplementation with N-3 long-chain polyunsaturated fatty acids or olive oil in men and women with renal disease induces differential changes in the DNA methylation of FADS2 and ELOVL5 in peripheral blood mononuclear cells. *PLoS One* **9**(10):e109896.

The Early Life Nutritional Environment, Epigenetics and Developmental Programming of Disease: Evidence from Animal Models

CHAPTER 3

Mark H. Vickers

Liggins Institute and Gravida,
National Centre for Growth and Development,
The University of Auckland,
Private Bag 92019, Victoria Street West,
Auckland 1142, New Zealand

Introduction

The nutritional environment in which the fetus or infant develops can influence the risk for developing metabolic and cardiovascular disorders in later life. The later onset of such diseases in response to earlier transient experiences suggests that developmental programming has an epigenetic component with epigenetic marks including DNA methylation, histone modifications or micro RNAs (miRNAs) providing a persistent memory of earlier nutritional state. As epigenetic regulation during development undergoes dynamic changes, the epigenome is labile and therefore allows adaptive responses to a changing environment, including an altered nutritional environment.[1] The mechanisms by which early environmental insults can have long-term effects on offspring remain largely undefined and evidence also exists, at least from animal models, that such epigenetic programming should be viewed as a transgenerational phenomenon.[2] These mechanisms include permanent structural changes to organs as a result of suboptimal levels of important factors during a critical developmental period, changes in gene expression arising as a consequence of epigenetic modifications and permanent changes in cellular aging.

Human studies are limited by a range of factors, including long generation times and quality of data records for retrospective studies, the accuracy and availability of methods to measure nutritional components in the one-carbon cycle, and investigation of tissue-specific effects. Most evidence for the epigenetic underpinnings of developmental programming has therefore been derived from a range of experimental animal models due to short generation times and scope for transgenerational studies, tissue-specific analysis and ability to implement intervention strategies that allow the observation of outcomes over the life course of the animal. Work in experimental models of programming has shown that a range of challenges during preconception, pregnancy and/or neonatal life can lead to changes in promoter methylation and thus directly or indirectly affect gene expression in pathways associated with a range of physiologic processes.[3–7] Methyl transfer or one-carbon metabolism is dependent upon dietary methyl donors and cofactors and research to date has therefore predominantly focused on dietary related changes in DNA methylation status.[8] although an increasing number of reports are now highlighting the role of nutrition manipulations on histone structure and function and miRNAs. Methylation of either DNA or histone proteins requires methyl donors from dietary folate and the presence of vitamins B6 and B12, choline, methionine, and a range of methyltransferases[9–11] and a framework is emerging integrating genomic methylation, histone modifications and miRNAs. Work on miRNAs has revealed a complex network of reciprocal interconnections: not only are they able to control gene expression at a post-transcriptional level, thus representing an important class of regulatory molecules, but they are also directly connected to the epigenetic machinery through a regulatory loop.[12] On one side, the expression of certain miRNAs is controlled by DNA methylation and chromatin modifications and, in turn, miRNAs can affect the methylation machinery and the expression of proteins involved in histone modifications — these mechanisms may therefore determine gene expression and the resultant phenotype. As an example, epigenetic silencing of miRNAs and miRNAs targeting histone deacetylases have been reported to play a role in cancer pathogenesis although their roles in the context of early life nutrition and developmental programming are not well-described. Global maternal under-nutrition (UN) has been shown to induce specific changes in cardiac miRNA profiles in adult rat offspring that are linked to cardiac structural changes, vascular dysfunction and hypertension.[13] Uteroplacental insufficiency has been shown to decrease lysine 36 trimethylation on histone 3 across the whole gene, with the greatest impact toward the 3' region of the gene. Interestingly, another rat model that uses maternal hyperglycemia as an early life event to induce adult cardiovascular disease similarly finds decreased lysine 36 trimethylation (H3Me3K36) of the hepatic insulin-like growth factor (IGF)-1 gene.[14] These data suggest that H3Me3K36 of the IGF-1 gene is sensitive to the glucose level of the prenatal environment, with resultant alterations in IGF-1 mRNA expression and ultimately vulnerability to adult onset insulin resistance.[14]

The extent of the window for the induction of epigenetic changes in key physiologic systems is not well-defined, but the period of developmental plasticity appears to extend from the periconceptional period into postnatal life.[6,15] Importantly, the phenotypic effects of epigenetic modifications during development may not manifest until later in life, especially if they affect genes modulating responses to later environmental challenges, such as post-weaning dietary challenges with high calorie diets i.e. a "second hit". Although most work has been undertaken in rodent models there are fundamental differences in imprinting control mechanisms between primate species and rodents at some imprinted domains. In the non-human primate model, correlations exist between miRNA expression and putative gene targets involved in developmental disorders and cardiovascular disease.[16] with implications for our understanding of the epigenetic programming process in humans and its influence on later disease risk.[17] However, human evidence remains largely unsubstantiated with the rodent data remaining the strongest argument for transgenerational epigenetic inheritance in humans.[18]

Maternal Under-nutrition

A number of studies using altered early life nutrition have now reported on changes in promoter methylation, histones or miRNAs and related effects on gene expression in pathways associated with a range of metabolic and cardiovascular disorders.[3–7,19] Studies with rodents have primarily utilized maternal under-nutrition (UN) (either global or low protein (LP)) or uterine artery ligation to induce intrauterine growth restricted (IUGR) offspring. Furthermore, although the role of macronutrients is clearly implicated in developmental programming,[20] maternal micronutrient levels are also of interest as they are essential for one-carbon metabolism and an imbalance in these nutrients can influence DNA methylation patterns in offspring. Organisms can fine-tune gene expression to achieve environmental adaptation via epigenetic alterations of histone markers of gene accessibility.[14] One example is that of IUGR induced via uteroplacental insufficiency leads to decreases in postnatal IGF1 mRNA variants, H3 acetylation and the gene elongation mark H3Me3K36.[14,21] Further work by Tosh et al. also showed that the pattern of early growth following IUGR (rapid versus delayed catch-up growth) in the rat leads to differential changes in hepatic IGF1 mRNA expression and histone H3K4 methylation.[11]

Maternal UN around the time of conception can induce changes in the expression of miRNAs in offspring which appear to play a role in the development of insulin resistance in later life.[22] Periconceptional UN in the sheep is associated with epigenetic changes in fetal hypothalamic POMC and GR genes, potentially resulting in altered energy balance regulation in the offspring. In the piglet model, reduced prenatal growth impairs remethylation capacity as well as ability to remove cystathionine and synthesize cysteine and taurine, which could have important implications on long-term health outcomes in IUGR offspring.[23] Feeding a maternal LP diet or creating uteroplacental insufficiency during pregnancy results in changes in DNA methylation which later manifest as endothelial dysfunction, hypertension and reduced renal glomeruli number.[5,24] Bogdarina and colleagues have shown that maternal

LP diet-induced hypertension in offspring is further associated with hypomethylation and increased expression of the angiotensin II type 1b (AT1b) receptor gene in the rat adrenal gland.[5] In the rat, altered promoter methylation and gene expression have been shown for the hepatic GR and the peroxisome proliferator-activated receptor (PPAR),[25,26] influencing carbohydrate and lipid metabolism.[27] Induction in offspring of altered epigenetic regulation of the hepatic GR promoter may be a consequence of reduced DNA methyltransferase (DNMT)-1 expression and changes in histones modifications.[26] Further, increasing the folic acid content of maternal or post-weaning diets can induce differential changes in phosphoenolpyruvate carboxykinase (PEPCK) mRNA expression and promoter methylation.[28] In addition to folate, evidence is growing for optimal dietary intake of choline for successful completion of fetal development.[29] Maternal choline supply during pregnancy in the rat modifies fetal histone and DNA methylation, suggesting that a concerted epigenomic mechanism contributes to the long term developmental effects of varied choline intake *in utero*.[30] Choline has been shown to be involved in the methylation of histone H3, expression of histone methyltransferases G9a (Kmt1c) and Suv39h1 (Kmt1a), and DNA methylation of their genes in rat fetal liver and brain.[30] The data on vitamin B12 in animal models are less clear. Vitamin B12 deficiency can result in hypomethylation as, along with folate, B12 is required for the synthesis of methionine and S-adenosyl methionine, the common methyl donor required for the maintenance of methylation patterns in DNA. A maternal LP diet alters miRNA profiles and mTOR expression in offspring influencing insulin secretion and glucose homeostasis with the glucose intolerance observed in adult offspring a result of a secretory defect in insulin rather than a reduced β-cell mass.[31] Pancreatic islets of LP offspring exhibited reduced mTOR expression and increased expression of a subset of miRNA and blockade of these miRNAs restore mTOR and insulin secretion to normal. A specific set of miRNAs plays a crucial role in pancreatic β-cell differentiation and is essential for the fine-tuning of insulin secretion and for compensatory β-cell mass expansion in response to insulin resistance.[32,33]

There are also important interactions between maternal phospholipid status and the one carbon cycle.[34] Inadequacy of long chain polyunsaturated fatty acids (LCPUFA's) containing phospholipids, one of the major methyl group acceptors in the one carbon metabolic pathway, may cause diversion of methyl groups toward DNA and eventually result in aberrant DNA methylation patterns.[35] These modified DNA methylation patterns lead to alterations in the expression of vital genes (e.g. angiogenic factors) and may thereby contribute to the dysregulation of angiogenesis/vasculogenesis further affecting placental development.[34] IUGR in rats is associated with decreased expression of pancreatic and duodenal homeobox factor-1 (PDX1), a key transcription factor regulating pancreatic development. Reduced PDX1 activity is associated with alterations in histone modifications.[36] Similar findings are observed for the glucose transporter GLUT4 in the muscle of IUGR rats.[37]

Mechanisms underlying epigenetic modification of tissue function resulting in a predisposition to altered programming of leptin and insulin signaling have been described previously.[38] Leptin has a 3-kb promoter region embedded within a CpG island and contains binding sites for known transcription factors including a glucocorticoid response element. Variations in DNA methylation of the leptin promoter in animal models depend on the degree of obesity[39,40] and the epigenetic regulation of leptin signaling pathways can be manipulated by nutrients and food compounds as detailed recently.[41] In addition to leptin, a further candidate re epigenetic regulation is that of the fat mass and obesity-associated (FTO) gene.[77–79] The potential programming effects of FTO in the developmental of obesity has been reviewed by Sebert and colleagues.[77] Experimental models in rodents and sheep (utilizing both under- and over-nutrition) have shown that the FTO gene may be a key target for nutritional programming. As an example, altered FTO methylation and gene regulation may provide a link between obesity-associated leptin resistance following IUGR and rapid postnatal weight gain.[80] However, as with many methylation data, although strong associations between FTO and BMI are consistently replicated in humans, the precise biological mechanisms by which FTO regulates weight gain remains unclear (e.g. does RNA demethylation impact on the control of energy balance?). In addition, different nutritional perturbations lead to differential effects on FTO regulation, e.g. placental expression of the obesity-associated gene FTO is reduced by fetal growth restriction but not by macrosomia in rats and humans.[81]

Maternal Obesity

In addition to maternal dietary restriction models, a number of reports now suggest that maternal obesity and early life over-nutrition can elicit epigenetic changes in offspring. Of note, phenotypic outcomes in offspring from both under-nutrition and over-nutrition experimental models display commonalities thus suggesting a "U"-shaped response curve to different early life nutritional perturbations. Whether the mechanisms are similar remains poorly defined. It also needs to be considered that obesity can in some cases represent a form of malnutrition due to deficiencies in key dietary nutrients thus may account for some of the phenotypic similarities observed.[42,43] Over-nutrition during the suckling period in the rat can lead to epigenetic modifications in key genes involved in the insulin signaling pathway in skeletal muscle and lead to later development of insulin resistance.[44] In the mouse, maternal obesity can induce epigenetic modifications that facilitate enhance adipogenic differentiation thereby programming for adiposity and metabolic dysfunction in later life.[45] Lesseur et al. have shown that maternal obesity and GDM are associated with tissue-specific changes in leptin promoter methylation in a gender-specific manner and may therefore have implications for later metabolic health of offspring.[46,47] Marco et al. showed that a maternal HFD induces hypermethylation of the hypothalamic POMC promoter and obesity in post-weaning rats.[48] A maternal HFD can also alter methylation and gene expression of dopamine and opioid-related genes, thus is a further potential mechanism for programming of appetite and preference for energy dense foods in postnatal life.[49]

A HFD and *in utero* exposure to obesity also disrupt circadian rhythm and lead to metabolic programming of liver in rat offspring.[50] These alterations included changes in PPARα transcription associated with epigenomic changes in H3K4me3 and H3K27me3 histone marks near the PPARα transcript start site. It has also been shown that maternal obesity interacts with an obesogenic post-weaning diet to promote the development of fatty liver disease via disruption of canonical metabolic rhythmicity gene expression in the liver.[51] This included hypermethylation of BMAL-1 (brain and muscle Arnt like-1) and Per2 promoter regions and altered 24-h rhythmicity of hepatic pro-inflammatory and fibrogenic mediators.

Offspring of mothers fed a moderate HFD diet display hepatic cell cycle inhibition and associated changes in gene expression and DNA methylation (Figure 1).[7] In particular, the cell cycle inhibitor cyclin-dependent kinase inhibitor (Cdkn1a, p21) is hypomethylated at specific CpG dinucleotides and hepatic mRNA expression increased in the liver of offspring of HFD mothers. Since Cdkn1a up-regulation has been associated with hepatocyte growth in pathologic states, these data may be suggestive of early hepatic

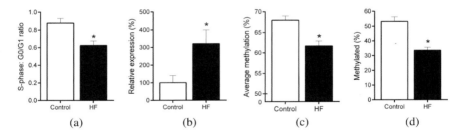

Figure 1. Offspring of mothers fed a HFD display cell cycle inhibition with associated changes in Cdkn1a gene expression and methylation at postnatal day 2 compared to offspring of control pregnancies. (a) Ratio of cells in a proliferative (S) phase relative to resting (G0/G1) phase; (b) hepatic gene expression of Cdkn1a; (c) average absolute hepatic DNA methylation across the entire CpG island; (d) absolute DNA methylation at a specific CpG unit shown to be significantly differentially methylated (position 7384643) between offspring of control and maternal HF pregnancies. Derived from Dudley et al.[7]

dysfunction in neonates born to HFD mothers. Exposure to maternal obesity therefore contributes to early perturbations in whole body and liver energy metabolism in offspring mediated in part by epigenetic effects. Early programming of mitochondrial dysfunction and a reduction in hepatic fatty acid oxidation may precede the development of detrimental obesity associated co-morbidities such as insulin resistance and hepatic steatosis in later life.[52-54] Several studies have reported alterations in miRNA levels in the islets of animal models of diabetes and in islets isolated from diabetic patients. However, many of the changes in miRNA expression observed in animal models of diabetes were not detected in the islets of diabetic patients and vice versa which may reflect fundamental differences in the experimental models.[55] Nutritional modulation of miRNAs may also underlie T2DM susceptibility in offspring. For example, a maternal HFD has been shown to alter hepatic miRNA profiles concomitant with changes in gene expression including IGF-2 which has been shown to be important for islet β-cell survival.[56]

Although studied less widely as an independent dietary co-factor, the effect of a high maternal salt intake, common in most obesogenic dietary environments, has also been examined in relation to epigenetic changes. Work by Ding *et al.* examined the influence of a high salt diet during pregnancy on heart development and DNA methylation in the fetal heart in relation to the subtype of angiotensin receptors. High maternal salt intake resulted in changes in a number of CpG sites compared to controls that were linked to the AT1b promoter in the fetal heart. Cardiac AT1 receptor protein in the adult offspring was also higher following exposure to a prenatal diet high in salt.[57]

Paternal Effects

The paternal influence on epigenetic alterations in offspring cannot be neglected and there is an increasing interest in the role of paternal effects in developmental programming. As the father solely transmits genetic and epigenetic factors to the oocyte, it has indeed been argued that the father may serve as a better model to explore epigenetic involvement in the setting of developmental programming.[58] Ng et al. reported that a chronic paternal HFD programmed β-cell dysfunction in female rat offspring and was paralleled by changes in DNA methylation profiles including hypomethylation of the Il13ra2 gene. This was the first report in mammals of nongenetic, intergenerational transmission of metabolic sequelae of a HFD from father to offspring.[59,60] Offspring of male mice fed a protein deplete diet and control fed females display an increased expression of genes involved in fat and cholesterol biosynthesis with increases in methylation in an enhancer for PPAR-α which could therefore regulate hepatic gene expression.[61] One of the most characterized epigenetically regulated loci, the paternally imprinted insulin-like growth factor-2 (IGF2) gene, is characterized by a labile methylation pattern dependent on the nutritional stimuli received by the growing organism during early life development.[62,63] Work by Waterland showed in post-weaning animals that the paternal allele of IGF2 DMR2 was hypermethylated in the kidneys of mice fed a control purified diet. This suggests that the nature of the post-weaning diet can permanently affect expression of IGF2, indicating that the childhood dietary environment could contribute to IGF2 loss of imprinting in humans.[63] Of note, the significant and independent association between paternal obesity and the offspring's methylation status suggests vulnerability of the developing sperm towards environmental insults. The acquired imprint instability may be carried onto the next generation and increase the risk for chronic diseases in adulthood.[64] Studies in F1 sperm have suggested a role for altered IGF2 and H19 expression in transmission of a phenotype to the F2 offspring.[65] However, not all studies reporting a paternal line transmission have reported epigenetic alterations in the F1 sperm.[66] Work by Radford et al. did not show any evidence that the epigenetic reprogramming of imprinting control regions in the germline was susceptible to nutritional restriction, thus suggesting that mechanisms other than direct germline transmission may be responsible.[67] The impact of paternal nutritional background in transgenerational inheritance has also been reported by Fullston et al.[68] where paternal obesity can initiate metabolic disturbances in two generations of mice although with incomplete penetrance to the F2 generation.

Epigenetics as a Tool to Inform on Interventions and Biomarker Development

Animal models have allowed investigation of possible intervention strategies to ameliorate or reverse the effects of developmental programming, including those detailed above involving maternal methyl donor supplementation. Leptin has been a focus of a number of programming-related studies with neonatal leptin treatment shown to reverse the effects of maternal under-nutrition in rodents[69,70] and in a piglet model of IUGR.[71] In the rodent, the effects of leptin treatment were sex-specific and changes in hepatic DNA methylation as a response to leptin treatment were directionally dependent upon prior maternal nutritional status.[72] Protective effects of leptin during the suckling period against later obesity may be associated with changes in promoter methylation of the hypothalamic POMC gene.[73] Mechanisms underlying epigenetic modification of tissue function resulting in a predisposition to altered programming of leptin and insulin signaling have been discussed previously.[38] It has been shown that leptin's promoter is subject to epigenetic programming, and leptin's expression can be modulated by DNA methylation.[74–76] In addition, activation of the leptin receptor also induces expression of suppressor of cytokine signaling-3 (SOCS-3). This protein inhibits further leptin signal transduction and also inhibits signaling by the insulin receptor. Altered SOCS-3 methylation may therefore have lasting effects on the leptin-insulin feedback loop and adversely impact developmental programming.[38] Yokomori et al. have shown that methylation of specific CpG sites and a methylation-sensitive protein could contribute to leptin gene expression during adipocyte differentiation in 3T3-L1 cells.[77] The same group has also shown that both methylation of specific CpG sites and a methylation-sensitive transcription factor contributes to GLUT4 gene regulation during preadipocyte to adipocyte differentiation.[78] In addition, differential

DNA methylation is observed in promoters of genes involved in glucose metabolism including GLUT4[78] and uncoupling protein 2 (UCP-2).[79] It addition to leptin, it has been shown that neonatal treatment with EX-4 increases histone acetylase activity and reverses epigenetic modifications that silence PDX1 in the IUGR rat.[80] Growth hormone (GH) treatment to neonatal rats born following maternal UN has also been shown to lead to changes in cardiac miRNA profiles in offspring.[13] Offspring of UN mothers are hypertensive and display cardiac hypertrophy and vascular dysfunction in adulthood.[81] GH treatment in the pre-weaning period leads to alterations in cardiac miRNAs in adulthood concomitant with normalization of blood pressure, cardiac mass and endothelial function. Moreover, the observed changes were specific to UN offspring and linked to the Let-7 miRNA family which have been reported to have a role in cardiac development.[82] In a mouse model of restricted maternal nutrition to induce a pre-eclampsia-type phenotype, pulmonary vascular dysfunction in offspring is associated with altered lung DNA methylation. Administration of histone deacetylase inhibitors (butyrate and trichostatin A) to offspring of diet restricted mice normalized pulmonary DNA methylation and pulmonary vascular function. Further, maternal administration of the nitroxide Tempol during dietary restriction prevented vascular dysfunction and dysmethylation in the offspring, thus further demonstrating the importance of epigenetic alterations in programming of later vascular function.[83,84] As noted previously, hypertension in offspring of LP-fed mothers is associated with changes in methylation and gene expression of the AT1b receptor[5] — supplementation to LP mothers with metyrapone, an 11β-hydroxylase inhibitor, during the first 14 days of pregnancy, normalized blood pressure, DNA methlyation and gene expression profiles in offspring.[85] Treatment with the constitutive androstane receptor ligand during pregnancy prevents insulin resistance in offspring from HFD diet-induced obese pregnant mice; effects mediated in part by abolishing the effect of a maternal HFD on epigenetic modifications of the genes encoding adiponectin and leptin in the offspring.[86] Taken together, these examples of intervention paradigms including leptin, GH and EX-4 highlight the importance of the early period of developmental plasticity as the key window for intervention to maximize effectiveness.[87]

Dietary supplementation with methyl donors as a strategy to ameliorate programming-related disorders has also been well-described. Dietary protein restriction of pregnant rats induces and folic acid supplementation prevents epigenetic modification of hepatic gene expression in the offspring.[25] Maternal choline supplementation can prevent increased adiposity and elevations in blood pressure in offspring following a maternal LP diet[88] and similar observations for restoration of normal blood pressure in offspring have been reported following maternal glycine supplementation.[89,90] Similarly, the obesogenic phenotype of offspring of dams fed a high multivitamin diet is prevented by a post-weaning high multivitamin or high folate diet although global methylation levels were similar across all groups.[91] High folate gestational and post-weaning diets alter hypothalamic feeding pathways via changes in DNA methylation in rat offspring (including POMC hypomethylation) with the obesogenic phenotype of offspring corrected by themselves being weaned onto a high folate diet.[91,92] This further highlights the post-weaning epigenetic plasticity of the hypothalamus and that *in utero* programming by vitamin modified gestational diets can be corrected by altering the vitamin content of the diet in offspring. However, it needs to be noted that rodent exposed *in utero* to a diet rich in folic acid and methyl donors can develop severe allergic airway disease, in part proposed to arise via increased methylation of Runx3.[93] Further, the importance of dietary methionine in programming of hypertension in the setting of a maternal LP diet has been highlighted. While one LP diet preparation containing methionine consistently produces hypertension in offspring, a further LP diet without methionine supplementation caused either no change or a slight reduction in blood pressure in offspring.[94,95]

Imbalances in maternal micronutrients can influence LCPUFA metabolism and global methylation in the rat placenta, mediated in part by a reduction in mRNA levels of methylene tetrahydrofolate reductase (MTHFR) and methionine synthase, and increased cystathionine b-synthase (CBS).[96] Maternal omega-3 supplementation can ameliorate many of these observed changes arising from maternal micronutrient imbalance through normalizing MTHFR and CBS levels. Maternal supplementation with conjugated linoleic acid (CLA) has been shown to improve maternal and offspring outcomes in the setting of maternal obesity[97,98] but the experimental data on CLA is conflicting and the role of epigenetic processes is not well-defined. In normal rodent pregnancies, CLA can reduce adipose tissue mass via apoptosis[99] and it has also been shown that CLA supplementation can lead to hypermethylation of the proximal specificity protein (Sp1) binding site that suppresses hypothalamic POMC in neonates and therefore may contribute to metabolic disorders in adults.[100] Post-weaning supplementation with adequate levels of selenium and folate in female offspring of mothers fed a HFD deficient in selenium and folate during gestation and lactation can alter global hepatic DNA methylation.[101] Further, in line with lifestyle modifications preventing mitochondrial alterations and metabolic disorders, exercise has also been shown to change DNA methylation of the promoter of PGC1a to favor gene expression responsible for mitochondrial biogenesis and function.[102] Maternal methyl donor supplementation reduces fatty liver and modifies the fatty acid synthase DNA methylation profile in rats fed an obesogenic diet.[103] Similarly, maternal methyl donor supplementation during lactation can prevent the hyperhomocysteinemia induced by a maternal high-fat-sucrose diet.[104] In addition to folic acid and vitamins B2, B6 and B12, other micronutrients such as vitamins A and C, iron, chromium, zinc, taurine and flavonoids play a role in developmental programming. A maternal high-zinc diet can attenuate intestinal inflammation by reducing DNA methylation and elevating H3K9 acetylation in the promoter of the anti-inflammatory protein A20.[105] The efficacy of the sulphonic acid taurine in ameliorating the adverse effects of both maternal under-nutrition and maternal obesity across a range of experimental animal models has also been reported[106–108] although the epigenetic basis of these observations is not well-defined. It is known that perinatal taurine depletion can increase oxidative stress and mediate blood pressure control throughout life and that early life taurine depletion leads to perinatal, epigenetic programming that impacts later physiological function.[109,110]

Aims currently pursued are the early identification of epigenetic biomarkers concerned in individuals' disease susceptibility and the description of protocols for tailored dietary treatments/advice to counterbalance adverse epigenomic events. These approaches may allow diagnosis and prognosis implementation and facilitate therapeutic strategies in a personalized "epigenomically modeled" manner to combat obesity and metabolic disorders.[111] It is also important to recognize sex-specific effects of interventions and there is no "one size fits all" intervention. This was highlighted in the leptin intervention studies whereby sex-specific responses to leptin treatment were observed[69,70] and also that leptin treatment to male offspring of normal pregnancies could induce an adverse metabolic phenotype in later life.[70] One caveat to note however is that hypothalamic development in large mammals is in contrast to the situation in rodents that are either altricial at birth or born immature.[112] Consequently, maturation of some key organ systems (including hypothalamic development) in the rodent is initiated prenatally but may not finalize until the second week of postnatal life.[113] Nutrient supply is therefore determined by maternal lactational capacity rather than being dependent on placental nutrient transfer capacity. However, despite the different developmental time windows across the range of experimental models used, both altricial and precocial species share several common programming mechanisms in offspring, particularly with regard to hypothalamic re-wiring.

Summary

Epigenetic alterations induced by suboptimal early life nutrition include altered DNA methylation, histone modifications, chromatin remodeling and/or regulatory feedback by miRNAs, all of which have the ability to modulate gene expression and promote the metabolic syndrome phenotype.[114] A wide range of nutritional interventions in pregnancy and lactation, including both undernutrition and maternal obesity, can lead to a range of metabolic disorders in offspring, which are mediated in part by epigenetic processes encompassing the chromatin information encrypted by DNA methylation patterns, histone covalent modifications and non-coding RNA or miRNA.[111] Most studies to date have addressed DNA methylation as the epigenetic mechanism with a focus on a range of micronutrients and one-carbon metabolism with histone modifications and miRNAs having less attention to date.[58] Methionine and folate are key components of one carbon metabolism, providing the methyl groups for numerous methyl transferase reactions via the ubiquitous methyl donor, s-adenosyl methionine. Methionine metabolism is responsive to nutrient intake, is regulated by several hormones and requires a number of vitamins as cofactors. The critical relationship between perturbations in the mother's methionine metabolism and its impact on fetal growth and development is now becoming evident via experimental animal models.[115] Evidence suggests that modifications of histone proteins turnover more rapidly than cell division, thereby bringing into question the involvement of histone modifications in transmission of epigenetic information (rapid histone turnover may serve to functionally separate chromatin domains and therefore prevent spread of histone states).[116,117]

It needs to be noted however, that most work to date is associative in nature and the extent to which epigenetic modifications may mediate the effects of developmental programming of altered energy balance remain unclear. For example, measuring DNA methylation in blood may provide little linkage with phenotype and therefore cannot be easily justified on a functional basis.[112] However, these epigenetic markers may have utility as predictive biomarkers of later disease risk (e.g. the link between cord blood promoter methylation of retinoid X receptor (rxr-α) and adiposity and bone mineral content in childhood).[118,119] That some traits appear to be resolved where others persist suggests that divergent mechanisms of transmission are involved and that those metabolic traits that do persist are capable of being transmitted via the male germline.[120] The evidence surrounding maternal folic acid and vitamin supplements, one carbon metabolism and altered DNA methylation patterns also raises the issue that more attention should be given to the potential long-term effects of such supplements on offspring (e.g. respiratory disorders or colitis) given that approximately 80% of women in the US take supplements during pregnancy.[58] The developmental window for the induction of epigenetic changes in key physiologic systems appears to extend from the periconceptional period into postnatal life.[6,15] Since the epigenetic processes are long-term and potentially reversible, once the mechanistic basis of the disease is understood, intervention and strategies aimed at reversal can be designed and implemented. As noted earlier, the phenotypic effects of epigenetic modifications during development may not manifest until adulthood, particularly when dependent upon interactions with the post-weaning environment such as that of a post-weaning energy dense diet. It also needs to be considered that, in addition to epigenetic effects, the contribution of the uterine tract environment or maternal adaptations to pregnancy may be critical to programming inheritance via the maternal line. Associations of maternal obesity with the neonatal epigenome have been shown to be stronger than associations with paternal obesity, supporting an intrauterine mechanism although weight gain during pregnancy was shown to have little effect.[121] Suboptimal nutrition *in utero* causes DNA damage and accelerated aging of the female reproductive tract[122] with the suggestion that developmental programming effects can be propagated through the maternal line *de novo* in generations beyond F2 as a consequence of development in a suboptimal intrauterine tract and not necessarily though directly transmitted epigenetic mechanisms.[2] Further, as the effects of age exacerbate the programmed metabolic phenotype, advancing maternal age may increase the likelihood of developmental programming effects being transmitted to future generations. However, many studies reported to date are to the F2 generation whereas true transgenerational transmission is the F3 generation and beyond where there is no exposure to the initial environmental challenge — however, data in F3 are limited and often variable depending on the model used.[2] In the meta-analysis by Aiken and Ozanne, more than half of studies completed through to F3 failed to show any effect.[2] A better understanding of the epigenetic basis of nutritionally mediated developmental programming and how these effects may be transmitted across generations is essential for the implementation of initiatives aimed at curbing the obesity and diabetes crisis. Thus, many key questions remain to be answered: How plastic is the system for intervention and what are the critical windows of development at which strategies should be targeted; what are the transgenerational effects and how many generations does it take to reverse epigenetic imprinting and can reliable markers be developed for disease prediction?[123] Further, many studies to date have failed to recognize the importance of sex-specific effects in programming and these differences need to be accounted for in future studies as they aid in the mechanistic understanding of how phenotypes evolve. Adopting a life course approach allows identification of phenotype and markers of risk earlier,[124] with the possibility of nutritional and other lifestyle interventions that have obvious implications for prevention of non-communicable diseases across generations.

References

1. H Jang, C Serra. (2014) Nutrition, epigenetics, and diseases. *Clin Nutr Res* **3**:1–8.
2. CE Aiken, SE Ozanne. (2014) Transgenerational developmental programming. *Hum Reprod Update* **20**:63–75.
3. RL Jirtle, MK Skinner. (2007) Environmental epigenomics and disease susceptibility. *Nat Rev Genet* **8**:253–262.
4. TD Pham, NK MacLennan, CT Chiu, GS Laksana, JL Hsu, RH Lane. (2003) Uteroplacental insufficiency increases apoptosis and alters p53 gene methylation in the full-term IUGR rat kidney. *Am J Physiol Regul Integr Comp Physiol* **285**:R962–970.
5. I Bogdarina, S Welham, PJ King, SP Burns, AJ Clark. (2007) Epigenetic modification of the renin-angiotensin system in the fetal programming of hypertension. *Circ Res* **100**:520–526.

6. IC Weaver, N Cervoni, FA Champagne, AC D'Alessio, S Sharma, JR Seckl, S Dymov, M Szyf, MJ Meaney. (2004) Epigenetic programming by maternal behavior. *Nat Neurosci* **7**:847–854.
7. KJ Dudley, DM Sloboda, KL Connor, J Beltrand, MH Vickers. (2011) Offspring of mothers fed a high fat diet display hepatic cell cycle inhibition and associated changes in gene expression and DNA methylation. *PLoS One* **6**:e21662.
8. B Delage, RH Dashwood. (2008) Dietary manipulation of histone structure and function. *Annu Rev Nutr* **28**:347–366.
9. AP Wolffe. (1998) Packaging principle: how DNA methylation and histone acetylation control the transcriptional activity of chromatin. *J Exp Zool* **282**:239–244.
10. PA Callinan, AP Feinberg. (2006) The emerging science of epigenomics. *Hum Mol Genet* **15**:R95–101.
11. DN Tosh, Q Fu, CW Callaway, RA McKnight, IC McMillen, MG Ross, RH Lane, M Desai. (2010) Epigenetics of programmed obesity: alteration in IUGR rat hepatic IGF1 mRNA expression and histone structure in rapid vs. delayed postnatal catch-up growth. *Am J Physiol Gastrointest Liver Physiol* **299**:G1023–1029.
12. Iorio MV, Piovan C, Croce CM. (2010) Interplay between microRNAs and the epigenetic machinery: an intricate network. *Biochim Biophys Acta* **1799**:694–701.
13. C Gray, M Li, CM Reynolds, MH Vickers. (2014) Let-7 miRNA profiles are associated with the reversal of left ventricular hypertrophy and hypertension in adult male offspring from mothers undernourished during pregnancy following pre-weaning growth hormone treatment. *Endocrinology* **155**:4808–4817.
14. EK Zinkhan, Q Fu, Y Wang, X Yu, CW Callaway, JL Segar, TD Scholz, RA McKnight, L Joss-Moore, RH Lane. (2012) Maternal hyperglycemia disrupts histone 3 lysine 36 trimethylation of the IGF-1 gene. *J Nutr Metab* **2012**:930364.
15. KD Sinclair, C Allegrucci, R Singh, DS Gardner, S Sebastian, J Bispham, A Thurston, JF Huntley, WD Rees, CA Maloney, RG Lea, J Craigon, TG McEvoy, LE Young. (2007) DNA methylation, insulin resistance, and blood pressure in offspring determined by maternal periconceptional B vitamin and methionine status. *Proc Natl Acad Sci USA* **104**:19351–19356.

16. A Maloyan, S Muralimanoharan, S Huffman, LA Cox, PW Nathanielsz, L Myatt, MJ Nijland. (2013) Identification and comparative analyses of myocardial miRNAs involved in the fetal response to maternal obesity. *Physiol Genomics* **45**:889–900.
17. CY Cheong, K Chng, S Ng, SB Chew, L Chan, AC Ferguson-Smith. (2015) Germline and somatic imprinting in the nonhuman primate highlights species differences in oocyte methylation. *Genome Res* **25**:611–623.
18. DK Morgan, E Whitelaw. (2008) The case for transgenerational epigenetic inheritance in humans. *Mamm Genome* **19**:394–397.
19. C Gicquel, A El-Osta, Y Le Bouc. (2008) Epigenetic regulation and fetal programming. *Best Pract Res Clin Endocrinol Metab* **22**:1–16.
20. IC McMillen, SM MacLaughlin, BS Muhlhausler, S Gentili, JL Duffield, JL Morrison. (2008) Developmental origins of adult health and disease: the role of periconceptional and foetal nutrition. *Basic Clin Pharmacol Toxicol* **102**:82–89.
21. Q Fu, RA McKnight, X Yu, L Wang, CW Callaway, RH Lane. (2004) Uteroplacental insufficiency induces site-specific changes in histone H3 covalent modifications and affects DNA-histone H3 positioning in day 0 IUGR rat liver. *Physiol Genomics* **20**:108–116.
22. S Lie, JL Morrison, O Williams-Wyss, CM Suter, DT Humphreys, SE Ozanne, S Zhang, SM Maclaughlin, DO Kleemann, SK Walker, CT Roberts, IC McMillen. (2014) Periconceptional undernutrition programs changes in insulin-signaling molecules and microRNAs in skeletal muscle in singleton and twin fetal sheep. *Biol Reprod* **90**:5.
23. DS MacKay, JD Brophy, LE McBreairty, RA McGowan, RF Bertolo. (2012) Intrauterine growth restriction leads to changes in sulfur amino acid metabolism, but not global DNA methylation, in Yucatan miniature piglets. *J Nutr Biochem* **23**:1121–1127.
24. J Morgado, B Sanches, R Anjos, C Coelho. (2015) Programming of essential hypertension: what pediatric cardiologists need to know. *Pediatr Cardiol* **36**:1327–1337.
25. KA Lillycrop, ES Phillips, AA Jackson, MA Hanson, GC Burdge. (2005) Dietary protein restriction of pregnant rats induces and folic acid supplementation prevents epigenetic modification of hepatic gene expression in the offspring. *J Nutr* **135**:1382–1386.

26. KA Lillycrop, JL Slater-Jefferies, MA Hanson, KM Godfrey, AA Jackson, GC Burdge. (2007) Induction of altered epigenetic regulation of the hepatic glucocorticoid receptor in the offspring of rats fed a protein-restricted diet during pregnancy suggests that reduced DNA methyltransferase-1 expression is involved in impaired DNA methylation and changes in histone modifications. *Br J Nutr* **97**:1064–1073.
27. GC Burdge, KA Lillycrop, AA Jackson, PD Gluckman, MA Hanson. (2007) The nature of the growth pattern and of the metabolic response to fasting in the rat are dependent upon the dietary protein and folic acid intakes of their pregnant dams and post-weaning fat consumption. *Br J Nutr* **99**:540–549.
28. SP Hoile, KA Lillycrop, LR Grenfell, MA Hanson, GC Burdge. (2012) Increasing the folic acid content of maternal or post-weaning diets induces differential changes in phosphoenolpyruvate carboxykinase mRNA expression and promoter methylation in rats. *Br J Nutr* **108**:852–857.
29. SH Zeisel. (2009) Importance of methyl donors during reproduction. *Am J Clin Nutr* **89**:673S–677S.
30. JM Davison, TJ Mellott, VP Kovacheva, JK Blusztajn. (2009) Gestational choline supply regulates methylation of histone H3, expression of histone methyltransferases G9a (Kmt1c) and Suv39h1 (Kmt1a), and DNA methylation of their genes in rat fetal liver and brain. *J Biol Chem* **284**:1982–1989.
31. EU Alejandro, B Gregg, T Wallen, D Kumusoglu, D Meister, A Chen, MJ Merrins, LS Satin, M Liu, P Arvan, E Bernal-Mizrachi. (2014) Maternal diet-induced microRNAs and mTOR underlie beta cell dysfunction in offspring. *J Clin Invest* **124**:4395–4410.
32. C Guay, E Roggli, V Nesca, C Jacovetti, R Regazzi. (2011) Diabetes mellitus, a microRNA-related disease? *Transl Res* **157**:253–264.
33. V Nesca, C Guay, C Jacovetti, V Menoud, ML Peyot, DR Laybutt, M Prentki, R Regazzi. (2013) Identification of particular groups of microRNAs that positively or negatively impact on beta cell function in obese models of type 2 diabetes. *Diabetologia* **56**:2203–2212.
34. V Khot, P Chavan-Gautam, S Joshi. (2015) Proposing interactions between maternal phospholipids and the one carbon cycle: A novel

mechanism influencing the risk for cardiovascular diseases in the offspring in later life. *Life Sci* **129**:16–21.
35. AA Khaire, AA Kale, SR Joshi. (2015) Maternal omega-3 fatty acids and micronutrients modulate fetal lipid metabolism: A review. *Prostaglandins Leukot Essent Fatty Acids* **98**:49–55.
36. JH Park, DA Stoffers, RD Nicholls, RA Simmons. (2008) Development of type 2 diabetes following intrauterine growth retardation in rats is associated with progressive epigenetic silencing of Pdx1. *J Clin Invest* **118**:2316–2324.
37. N Raychaudhuri, S Raychaudhuri, M Thamotharan, SU Devaskar. (2008) Histone code modifications repress glucose transporter 4 expression in the intrauterine growth-restricted offspring. *J Biol Chem* **283**:13611–13626.
38. MJ Holness, MC Sugden. (2006) Epigenetic regulation of metabolism in children born small for gestational age. *Curr Opin Clin Nutr Metab Care* **9**:482–488.
39. W Shen, C Wang, L Xia, C Fan, H Dong, RJ Deckelbaum, K Qi. (2014) Epigenetic modification of the leptin promoter in diet-induced obese mice and the effects of N-3 polyunsaturated fatty acids. *Sci Rep* **4**:5282.
40. AB Crujeiras, MC Carreira, B Cabia, S Andrade, M Amil, FF Casanueva. (2015) Leptin resistance in obesity: an epigenetic landscape. *Life Sci* **140**:57–63.
41. P Haggarty. (2013) Epigenetic consequences of a changing human diet. *Proc Nutr Soc* **72**:363–371.
42. JE Kimmons, HM Blanck, BC Tohill, J Zhang, LK Khan. (2006) Associations between body mass index and the prevalence of low micronutrient levels among US adults. *MedGenMed* **8**:59.
43. JC Kerns, C Arundel, LS Chawla. (2015) Thiamin deficiency in people with obesity. *Adv Nutr* **6**:147–153.
44. HW Liu, S Mahmood, M Srinivasan, DJ Smiraglia, MS Patel. (2013) Developmental programming in skeletal muscle in response to overnourishment in the immediate postnatal life in rats. *J Nutr Biochem* **24**:1859–1869.
45. QY Yang, JF Liang, CJ Rogers, JX Zhao, MJ Zhu, M Du. (2013) Maternal obesity induces epigenetic modifications to facilitate

Zfp423 expression and enhance adipogenic differentiation in fetal mice. *Diabetes* **62**:3727–3735.
46. C Lesseur, DA Armstrong, AG Paquette, DC Koestler, JF Padbury, CJ Marsit. (2013) Tissue-specific leptin promoter DNA methylation is associated with maternal and infant perinatal factors. *Mol Cell Endocrinol* **381**:160–167.
47. C Lesseur, DA Armstrong, AG Paquette, Z Li, JF Padbury, CJ Marsit. (2014) Maternal obesity and gestational diabetes are associated with placental leptin DNA methylation. *Am J Obstet Gynecol* **211**:654 e651–659.
48. A Marco, T Kisliouk, A Weller, N Meiri. (2013) High fat diet induces hypermethylation of the hypothalamic pomc promoter and obesity in post-weaning rats. *Psychoneuroendocrinology* **38**:2844–2853.
49. Z Vucetic, J Kimmel, K Totoki, E Hollenbeck, TM Reyes. (2010) Maternal high-fat diet alters methylation and gene expression of dopamine and opioid-related genes. *Endocrinology* **151**:4756–4764.
50. SJ Borengasser, P Kang, J Faske, H Gomez–Acevedo, ML Blackburn, TM Badger, K Shankar. (2014) High fat diet and in utero exposure to maternal obesity disrupts circadian rhythm and leads to metabolic programming of liver in rat offspring. *PLoS One* **9**:e84209.
51. A Mouralidarane, J Soeda, D Sugden, A Bocianowska, R Carter, S Ray, R Saraswati, P Cordero, M Novelli, G Fusai, M Vinciguerra, L Poston, PD Taylor, JA Oben. (2015) Maternal obesity programs offspring non-alcoholic fatty liver disease through disruption of 24-h rhythms in mice. *Int J Obes (Lond)* **39**:1339–1348.
52. DE Brumbaugh, JE Friedman. (2014) Developmental origins of nonalcoholic fatty liver disease. *Pediatr Res* **75**:140–147.
53. M Kjaergaard, C Nilsson, A Rosendal, MO Nielsen, K Raun. (2014) Maternal chocolate and sucrose soft drink intake induces hepatic steatosis in rat offspring associated with altered lipid gene expression profile. *Acta Physiol (Oxf)* **210**:142–153.
54. TJ Pereira, MA Fonseca, KE Campbell, BL Moyce, LK Cole, GM Hatch, CA Doucette, J Klein, M Aliani, VW Dolinsky. (2015) Maternal obesity characterized by gestational diabetes increases the susceptibility of rat offspring to hepatic steatosis via a disrupted liver metabolome. *J Physiol* **593**:3181–3197.

55. C Guay, R Regazzi. (2014) Role of islet microRNAs in diabetes: which model for which question? *Diabetologia* **58**:456–463.
56. J Zhang, F Zhang, X Didelot, KD Bruce, FR Cagampang, M Vatish, M Hanson, H Lehnert, A Ceriello, CD Byrne. (2009) Maternal high fat diet during pregnancy and lactation alters hepatic expression of insulin like growth factor-2 and key microRNAs in the adult offspring. *BMC Genomics* **10**:478.
57. Y Ding, J Lv, C Mao, H Zhang, A Wang, L Zhu, H Zhu, Z Xu. (2010) High-salt diet during pregnancy and angiotensin-related cardiac changes. *J Hypertens* **28**:1290–1297.
58. K Vanhees, IG Vonhogen, FJ van Schooten, RW Godschalk. (2014) You are what you eat, and so are your children: the impact of micronutrients on the epigenetic programming of offspring. *Cell Mol Life Sci* **71**:271–285.
59. SF Ng, RC Lin, DR Laybutt, R Barres, JA Owens, MJ Morris. (2010) Chronic high-fat diet in fathers programs beta-cell dysfunction in female rat offspring. *Nature* **467**:963–966.
60. SF Ng, RC Lin, CA Maloney, NA Youngson, JA Owens, MJ Morris. (2014) Paternal high-fat diet consumption induces common changes in the transcriptomes of retroperitoneal adipose and pancreatic islet tissues in female rat offspring. *FASEB J* **28**:1830–1841.
61. BR Carone, L Fauquier, N Habib, JM Shea, CE Hart, R Li, C Bock, C Li, H Gu, PD Zamore, A Meissner, Z Weng, HA Hofmann, N Friedman, OJ Rando. (2010) Paternally induced transgenerational environmental reprogramming of metabolic gene expression in mammals. *Cell* **143**:1084–1096.
62. SK Murphy, RL Jirtle. (2003) Imprinting evolution and the price of silence. *Bioessays* **25**:577–588.
63. RA Waterland, JR Lin, CA Smith, RL Jirtle. (2006) Post-weaning diet affects genomic imprinting at the insulin-like growth factor 2 (Igf2) locus. *Hum Mol Genet* **15**:705–716.
64. A Soubry, SK Murphy, F Wang, Z Huang, AC Vidal, BF Fuemmeler, J Kurtzberg, A Murtha, RL Jirtle, JM Schildkraut, C Hoyo. (2015) Newborns of obese parents have altered DNA methylation patterns at imprinted genes. *Int J Obes (Lond)* **39**:650–657.

65. GL Ding, FF Wang, J Shu, S Tian, Y Jiang, D Zhang, N Wang, Q Luo, Y Zhang, F Jin, PC Leung, JZ Sheng, HF Huang. (2012) Transgenerational glucose intolerance with Igf2/H19 epigenetic alterations in mouse islet induced by intrauterine hyperglycemia. *Diabetes* **61**:1133–1142.
66. AJ Drake, L Liu, D Kerrigan, RR Meehan, JR Seckl. (2011) Multigenerational programming in the glucocorticoid programmed rat is associated with generation-specific and parent of origin effects. *Epigenetics* **6**:1334–1343.
67. EJ Radford, E Isganaitis, J Jimenez-Chillaron, J Schroeder, M Molla, S Andrews, N Didier, M Charalambous, K McEwen, G Marazzi, D Sassoon, ME Patti, AC Ferguson-Smith. (2012) An unbiased assessment of the role of imprinted genes in an intergenerational model of developmental programming. *PLoS Genet* **8**:e1002605.
68. T Fullston, EM Ohlsson Teague, NO Palmer, MJ DeBlasio, M Mitchell, M Corbett, CG Print, JA Owens, M Lane. (2013) Paternal obesity initiates metabolic disturbances in two generations of mice with incomplete penetrance to the F2 generation and alters the transcriptional profile of testis and sperm microRNA content. *FASEB J* **27**:4226–4243.
69. MH Vickers, PD Gluckman, AH Coveny, PL Hofman, WS Cutfield, A Gertler, BH Breier, M Harris. (2005) Neonatal leptin treatment reverses developmental programming. *Endocrinology* **146**: 4211–4216.
70. MH Vickers, PD Gluckman, AH Coveny, PL Hofman, WS Cutfield, A Gertler, BH Breier, M Harris. (2008) The effect of neonatal leptin treatment on postnatal weight gain in male rats is dependent on maternal nutritional status during pregnancy. *Endocrinology* **149**:1906–1913.
71. L Attig, J Djiane, A Gertler, O Rampin, T Larcher, S Boukthir, PM Anton, JY Madec, I Gourdou, L Abdennebi-Najar. (2008) Study of hypothalamic leptin receptor expression in low-birth-weight piglets and effects of leptin supplementation on neonatal growth and development. *Am J Physiol Endo Metab* **295**: E1117–1125.

72. PD Gluckman, KA Lillycrop, MH Vickers, AB Pleasants, ES Phillips, AS Beedle, GC Burdge, MA Hanson. (2007) Metabolic plasticity during mammalian development is directionally dependent on early nutritional status. *Proc Natl Acad Sci USA* **104**:12796–12800.
73. M Palou, C Pico, JA McKay, J Sanchez, T Priego, JC Mathers, A Palou. (2011) Protective effects of leptin during the suckling period against later obesity may be associated with changes in promoter methylation of the hypothalamic pro-opiomelanocortin gene. *Br J Nutr* **106**:769–778.
74. D Iliopoulos, KN Malizos, A Tsezou. (2007) Epigenetic regulation of leptin affects MMP-13 expression in osteoarthritic chondrocytes: possible molecular target for osteoarthritis therapeutic intervention. *Ann Rheum Dis* **66**:1616–1621.
75. I Melzner, V Scott, K Dorsch, P Fischer, M Wabitsch, S Bruderlein, C Hasel, P Moller. (2002) Leptin gene expression in human preadipocytes is switched on by maturation-induced demethylation of distinct CpGs in its proximal promoter. *J Biol Chem* **277**:45420–45427.
76. R Stoger. (2006) In vivo methylation patterns of the leptin promoter in human and mouse. *Epigenetics* **1**:155–162.
77. N Yokomori, M Tawata, T Onaya. (2002) DNA demethylation modulates mouse leptin promoter activity during the differentiation of 3T3-L1 cells. *Diabetologia* **45**:140–148.
78. N Yokomori, M Tawata, T Onaya. (1999) DNA demethylation during the differentiation of 3T3-L1 cells affects the expression of the mouse GLUT4 gene. *Diabetes* **48**:685–690.
79. MV Carretero, L Torres, U Latasa, ER Garcia-Trevijano, J Prieto, JM Mato, MA Avila. (1998) Transformed but not normal hepatocytes express UCP2. *FEBS Lett* **439**:55–58.
80. SE Pinney, LJ Jaeckle Santos, Y Han, DA Stoffers, RA Simmons. (2011) Exendin-4 increases histone acetylase activity and reverses epigenetic modifications that silence Pdx1 in the intrauterine growth retarded rat. *Diabetologia* **54**:2606–2614.
81. C Gray, M Li, CM Reynolds, MH Vickers. (2013) Pre-weaning growth hormone treatment reverses hypertension and endothelial dysfunction in adult male offspring of mothers undernourished during pregnancy. *PLoS One* **8**:e53505.

82. KT Kuppusamy, DC Jones, H Sperber, A Madan, KA Fischer, ML Rodriguez, L Pabon, WZ Zhu, NL Tulloch, X Yang, NJ Sniadecki, MA Laflamme, WL Ruzzo, CE Murry, H Ruohola-Baker. (2015) Let-7 family of microRNA is required for maturation and adult-like metabolism in stem cell-derived cardiomyocytes. *Proc Natl Acad Sci USA* **112**:E2785–2794.
83. U Scherrer, SF Rimoldi, C Sartori, FH Messerli, E Rexhaj. (2015) Fetal programming and epigenetic mechanisms in arterial hypertension. *Curr Opin Cardiol* **30**:393–397.
84. E Rexhaj, J Bloch, PY Jayet, SF Rimoldi, D Dessen, C Mathieu, JF Tolsa, P Nicod, U Scherrer, C Sartori. (2011) Fetal programming of pulmonary vascular dysfunction in mice: role of epigenetic mechanisms. *Am J Physiol Heart Circ Physiol* **301**:H247–252.
85. I Bogdarina, A Haase, S Langley-Evans, AJ Clark. (2010) Glucocorticoid effects on the programming of AT1b angiotensin receptor gene methylation and expression in the rat. *PLoS One* **5**:e9237.
86. H Masuyama, Y Hiramatsu. (2012) Treatment with constitutive androstane receptor ligand during pregnancy prevents insulin resistance in offspring from high-fat diet-induced obese pregnant mice. *Am J Physiol Endocrinol Metab* **303**:E293–300.
87. KM Godfrey, PD Gluckman, MA Hanson. (2010) Developmental origins of metabolic disease: life course and intergenerational perspectives. *Trends in Endo Metab* **21**:199–205.
88. S Bai, D Briggs, MH Vickers. (2012) Increased systolic blood pressure in rat offspring following a maternal low-protein diet is normalized by maternal dietary choline supplementation. *J Develop Origins of Health and Disease* **3**:1–8.
89. L Brawley, C Torrens, FW Anthony, S Itoh, T Wheeler, AA Jackson, GF Clough, L Poston, MA Hanson. (2004) Glycine rectifies vascular dysfunction induced by dietary protein imbalance during pregnancy. *J Physiol* **554**:497–504.
90. AA Jackson, RL Dunn, MC Marchand, SC Langley-Evans. (2002) Increased systolic blood pressure in rats induced by a maternal low-protein diet is reversed by dietary supplementation with glycine. *Clin Sci (Lond)* **103**:633–639.

91. CE Cho, D Sanchez-Hernandez, SA Reza-Lopez, PS Huot, YI Kim, GH Anderson. (2013) Obesogenic phenotype of offspring of dams fed a high multivitamin diet is prevented by a post-weaning high multivitamin or high folate diet. *Int J Obes (Lond)* **37**:1177–1182.
92. CE Cho, D Sanchez-Hernandez, SA Reza-Lopez, PS Huot, YI Kim, GH Anderson. (2013) High folate gestational and post-weaning diets alter hypothalamic feeding pathways by DNA methylation in Wistar rat offspring. *Epigenetics* **8**:710–719.
93. JW Hollingsworth, S Maruoka, K Boon, S Garantziotis, Z Li, J Tomfohr, N Bailey, EN Potts, G Whitehead, DM Brass, DA Schwartz. (2008) In utero supplementation with methyl donors enhances allergic airway disease in mice. *J Clin Invest* **118**:3462–3469.
94. SC Langley, AA Jackson. (1994) Increased systolic blood pressure in adult rats induced by fetal exposure to maternal low protein diets. *Clin Sci (Lond)* **86**:217–222; discussion 121.
95. SC Langley-Evans. (2000) Critical differences between two low protein diet protocols in the programming of hypertension in the rat. *Int J Food Sci Nutr* **51**:11–17.
96. V Khot, A Kale, A Joshi, P Chavan-Gautam, S Joshi. (2014) Expression of genes encoding enzymes involved in the one carbon cycle in rat placenta is determined by maternal micronutrients (folic acid, vitamin B12) and omega-3 fatty acids. *Biomed Res Int* **2014**:613078.
97. C Gray, MH Vickers, SA Segovia, XD Zhang, CM Reynolds. (2015) A maternal high fat diet programmes endothelial function and cardiovascular status in adult male offspring independent of body weight, which is reversed by maternal conjugated linoleic acid (CLA) supplementation. *PLoS One* **10**:e0115994.
98. CM Reynolds, SA Segovia, XD Zhang, C Gray, MH Vickers. (2015) Conjugated linoleic acid supplementation during pregnancy and lactation reduces maternal high-fat-diet-induced programming of early-onset puberty and hyperlipidemia in female rat offspring. *Biol Reprod* **92**:40.
99. N Tsuboyama-Kasaoka, M Takahashi, K Tanemura, HJ Kim, T Tange, H Okuyama, M Kasai, S Ikemoto, O Ezaki. (2000) Conjugated

linoleic acid supplementation reduces adipose tissue by apoptosis and develops lipodystrophy in mice. *Diabetes* **49**:1534–1542.

100. X Zhang, R Yang, Y Jia, D Cai, B Zhou, X Qu, H Han, L Xu, L Wang, Y Yao, G Yang. (2014) Hypermethylation of Sp1 binding site suppresses hypothalamic POMC in neonates and may contribute to metabolic disorders in adults: impact of maternal dietary CLAs. *Diabetes* **63**:1475–1487.

101. EN Bermingham, SA Bassett, W Young, NC Roy, WC McNabb, JM Cooney, DT Brewster, WA Laing, MP Barnett. (2013) Post-weaning selenium and folate supplementation affects gene and protein expression and global DNA methylation in mice fed high-fat diets. *BMC Med Genomics* **6**:7.

102. Z Cheng, FA Almeida. (2014) Mitochondrial alteration in type 2 diabetes and obesity: an epigenetic link. *Cell Cycle* **13**:890–897.

103. P Cordero, AM Gomez-Uriz, J Campion, FI Milagro, JA Martinez. (2013) Dietary supplementation with methyl donors reduces fatty liver and modifies the fatty acid synthase DNA methylation profile in rats fed an obesogenic diet. *Genes Nutr* **8**:105–113.

104. P Cordero, FI Milagro, J Campion, JA Martinez. (2013) Maternal methyl donors supplementation during lactation prevents the hyperhomocysteinemia induced by a high-fat-sucrose intake by dams. *Int J Mol Sci* **14**:24422–24437.

105. C Li, S Guo, J Gao, Y Guo, E Du, Z Lv, B Zhang. (2015) Maternal high-zinc diet attenuates intestinal inflammation by reducing DNA methylation and elevating H3K9 acetylation in the A20 promoter of offspring chicks. *J Nutr Biochem* **26**:173–183.

106. M Li, CM Reynolds, DM Sloboda, C Gray, MH Vickers. (2013) Effects of taurine supplementation on hepatic markers of inflammation and lipid metabolism in mothers and offspring in the setting of maternal obesity. *PLoS One* **8**:e76961.

107. M Li, CM Reynolds, DM Sloboda, C Gray, MH Vickers. (2015) Maternal taurine supplementation attenuates maternal fructose-induced metabolic and inflammatory dysregulation and partially reverses adverse metabolic programming in offspring. *J Nutr Biochem* **26**:267–276.

108. S Boujendar, E Arany, D Hill, C Remacle, B Reusens. (2003) Taurine supplementation of a low protein diet fed to rat dams normalizes the vascularization of the fetal endocrine pancreas. *J Nutr* **133**: 2820–2825.
109. W Lerdweeraphon, JM Wyss, T Boonmars, S Roysommuti. (2013) Perinatal taurine exposure affects adult oxidative stress. *Am J Physiol Regul Integr Comp Physiol* **305**:R95–97.
110. S Roysommuti, JM Wyss. (2014) Perinatal taurine exposure affects adult arterial pressure control. *Amino Acids* **46**:57–72.
111. JA Martinez, P Cordero, J Campion, FI Milagro. (2012) Interplay of early-life nutritional programming on obesity, inflammation and epigenetic outcomes. *Proc Nutr Soc* **71**:276–283.
112. S Ojha, HP Fainberg, S Sebert, H Budge, ME Symonds. (2015) Maternal health and eating habits: metabolic consequences and impact on child health. *Trends Mol Med* **21**:126–133.
113. SG Bouret, RB Simerly (2006) Developmental programming of hypothalamic feeding circuits. *Clin Genet* **70**:295–301.
114. M Desai, JK Jellyman, MG Ross. (2015) Epigenomics, gestational programming and risk of metabolic syndrome. *Int J Obes (Lond)* **39**:633–641.
115. SC Kalhan, SE Marczewski. (2012) Methionine, homocysteine, one carbon metabolism and fetal growth. *Rev Endocr Metab Disord* **13**:109–119.
116. RB Deal, JG Henikoff, S Henikoff. (2010) Genome-wide kinetics of nucleosome turnover determined by metabolic labeling of histones. *Science* **328**:1161–1164.
117. MF Dion, T Kaplan, M Kim, S Buratowski, N Friedman, OJ Rando. (2007) Dynamics of replication-independent histone turnover in budding yeast. *Science* **315**:1405–1408.
118. KM Godfrey, A Sheppard, PD Gluckman, KA Lillycrop, GC Burdge, C McLean, J Rodford, JL Slater-Jefferies, E Garratt, SR Crozier, BS Emerald, CR Gale, HM Inskip, C Cooper, MA Hanson. (2011) Epigenetic gene promoter methylation at birth is associated with child's later adiposity. *Diabetes* **60**:1528–1534.
119. NC Harvey, A Sheppard, KM Godfrey, C McLean, E Garratt, G Ntani, L Davies, R Murray, HM Inskip, PD Gluckman, MA Hanson,

KA Lillycrop, C Cooper. (2014) Childhood bone mineral content is associated with methylation status of the RXRA promoter at birth. *J Bone Miner Res* **29**:600–607.

120. GA Dunn, TL Bale. (2011) Maternal high-fat diet effects on third-generation female body size via the paternal lineage. *Endocrinology* **152**:2228–2236.

121. GC Sharp, DA Lawlor, RC Richmond, A Fraser, A Simpkin, M Suderman, HA Shihab, O Lyttleton, W McArdle, SM Ring, TR Gaunt, G Davey Smith, CL Relton. (2015) Maternal pre-pregnancy BMI and gestational weight gain, offspring DNA methylation and later offspring adiposity: findings from the Avon longitudinal study of parents and children. *Int J Epidemiol* **44**: 1288–1304.

122. CE Aiken, JL Tarry-Adkins, SE Ozanne. (2013) Suboptimal nutrition in utero causes DNA damage and accelerated aging of the female reproductive tract. *FASEB J* **27**:3959–3965.

123. FI Milagro, ML Mansego, C De Miguel, JA Martinez. (2013) Dietary factors, epigenetic modifications and obesity outcomes: progresses and perspectives. *Mol Aspects Med* **34**:782–812.

124. KM Godfrey, PD Gluckman, MA Hanson. (2010) Developmental origins of metabolic disease: life course and intergenerational perspectives. *Trends Endocrinol Metab* **21**:199–205.

Lipids and Epigenetics

CHAPTER 4

Graham C. Burdge

Academic Unit of Human Development and Health,
Faculty of Medicine, University of Southampton,
IDS Building (MP 887), Southampton Genral Hospital
Tremona Road, Southampton, SO16 6YD, UK

Introduction

Epigenetics refers to a group of closely interrelated processes that regulate transcription. These are methylation at the 5′ position of cytosine bases in CpG dinucleotide pairs, covalent modifications of histones (principally acetylation and methylation of lysine residues) and the activities of non-coding RNA species.[1] Epigenetic processes, in particular DNA methylation, are critical for the induction and maintenance of cellular phenotype and for mediating parental influence on growth and development.[1] Parentally imprinted marks are induced during gamete formation and are retained during genome-wide demethylation that produces pluripotent stem cells after fertilization. Cell-type-specific epigenetic marks are induced during embryogenesis. Both imprinted and cell-type-specific epigenetic marks are retained throughout the life course. However, covalent modifications of histones may vary more dynamically than DNA methylation marks. Furthermore, a proportion of CpG loci retain plasticity throughout the life course[2] and can be modified by environmental inputs including diet. It is not known whether such plasticity is restricted to specific periods of the life course, such as development and aging, or whether there is a persistent susceptibility to environmental inputs throughout the life course that increases during specific periods. One implication of epigenetic plasticity is that although variations in the epigenetic control of genes involved in nutrient metabolism may contribute to individual differences in nutrient requirements, dietary choices may modify the epigenome and hence influence the levels of nutrient intakes that are needed for health.[3] The effects of nutrients that are involved in 1-carbon metabolism and protein, possibly acting via the provision of amino acids that function as methyl donors, are well-established.[4] However, there is increasing evidence that lipids can also modify epigenetic processes.

Dietary Lipids and the Epigenome

This chapter will focus specifically on the effects of dietary fatty acids on epigenetic processes. The effects of obesity on the epigenome will not be considered in this chapter given the potential influence of factors in addition to increased fat mass such as changes in the levels of hormones and inflammatory mediators that may mask any specific effects of fatty acids.

The Effect of Treatment with Fatty Acids on Cultured Cells

Short and long chain fatty acids have been shown to induce changes in DNA methylation and in covalent histone modifications in several different types of cell in culture. The short chain fatty acid butyric acid is obtained from dairy products[5] via breakdown of dietary fiber by the anaerobic fermentation of colonic bacteria.[6] Butyrate has been shown to be a potent inhibitor of histone deacetylases[7] and, for example, to modulate neuronal differentiation,[8] skeletal development[9] and capacity for differentiation of mesenchymal stem cells[10] by increasing the level of histone acetylation and thus facilitating a transcriptionally permissive state. Such effects appear to affect specific groups of genes, primarily those associated with cell differentiation and cell cycle,[11,12] and to be targeted at particular lysine resides. For example, in MCF-7 cells butyrate treatment was found to increase acetylation of H3K9, but not H4K16.[13] Furthermore, inhibition of histone deactylase (HDAC)-1 by butyrate has been shown to increase histone acetylation and induced rapid DNA demethylation of retinoic acid receptor-β_2 leading to an overall transcriptionally permissive state.[14] One study has found that butyrate can increase gene expression and replication of non-genomically integrated HIV-1.[15] Valproic acid has also been found to modify neuronal differentiation by a mechanism involving hyperacetylation of histone H3.[16]

Fewer studies have reported the effects of long chain fatty acids on epigenetic marks in cultured cells. The saturated fatty acid palmitic acid has been shown to induce a relatively small increase in global DNA methylation (approximately 1%) in primary pancreatic islet cells.[17] These changes were localized primarily to shelf and shore regions of the 5′ regulatory regions, but not within CpG islands. Although altered mRNA expression was reported for a number of genes that have been associated with type 2 diabetes mellitus and obesity, no data were shown that linked altered methylation directly to changes in the transcriptome. Thus it remains uncertain whether a change of 1% in global methylation is sufficient to alter transcription. Treatment of myotubes derived from skeletal muscle cells from lean and obese individuals with a mixture of palmitic and the monounsaturated fatty acid oleic acid (1:1; 250 μmol/l) induced small (<5%) changes in the methylation of individual CpG loci in the PPARδ coding region that differed between individuals with different body-mass-index.[18]

Treatment of M17 neuoblastoma cells with the polyunsaturated fatty acid docosahexaenoic acid (DHA) (10 μmol/l) for 48 hours induced increased global H3K9 acetylation and decreased global dimethyl H3K4, dimethyl H3K9, dimethyl H3K27, dimethyl H3K36 and dimethyl H3K79 levels.[19] This was accompanied by decreased protein expression of histone deacetylases 1, 2 and 3. Overall, treatment with DHA tended to induce histone changes that are consistent with a transcriptionally permissive state in manner similar to butyrate, although the effect of DHA on HDAC activity was not tested. Unfortunately, this study did not test whether such effects were specific to DHA or whether they were dependent on DHA concentration. Treatment of U937 leukaemic cells with eicosapentaenoic acid (EPA) (μmol/l) increased mRNA expression of CCAAT/enhancer-binding proteins C/EBP-β and C/EBP-δ, PU.1 and c-Jun.[20] This was accompanied by demethylation of a single CpG locus the C/EBP-δ promoter compared to untreated cells or cells treated with oleic acid. Furthermore, treatment of U937 cells with EPA also induced almost complete demethylation of the H-Ras intron 1 CpG island which was accompanied by increased H-Ras protein expression compared to untreated cells or cells treated with oleic acid.[21]

Together, these findings show that treatment of cultured cells with fatty acids can induce changes in the epigenetic processes in a manner contingent on the fatty acid and which exhibits specificity in terms of target epigenetic mark and the genes that are affected.

Animal Models of Altered Maternal Fat Intake

Although developing mammals have some capacity for synthesis of saturated and monounsaturated fatty acids, the majority of their requirements for fatty acids, in particular polyunsaturated fatty acids, are met by the maternal diet and by adaptions to maternal metabolism. The period in development in which specific fatty acids are assimilated by the fetus differs markedly between species. For example, animals that are born at a relatively early stage of development such as rodents tend to accumulate fatty acids in adipose tissue and in specific organs such as the brain[22] after birth. In these species lactation is the primary route of fatty acids supply. Other species, including humans and guinea pigs, tend to develop substantial adipose tissue reserves[23] and incorporate fatty acids into specific organs, for example the brain,[24] before birth. In these animals, both placenta fatty acid transport and lactation play important roles in providing fatty acids during development. However, whether fatty acids are supplied primarily through lactation or by a combination of placental transfer and lactation, pregnant females undergo metabolic adaptations to facilitate supply of fatty acids either via the placenta or milk including hyperlipidaemia, insulin resistance in late gestation[25] and changes in phospholipid and polyunsaturated fatty acid metabolism.[26,27] One implication is that dietary interventions during pregnancy can be modified by maternal metabolism and the stage of development of the offspring and the species of animal. These factors have implications for interpreting studies of the impact of maternal dietary fat on the epigenome of the offspring.

The reports published to date of the effect of maternal dietary fat on the epigenome of the offspring have focused mainly on studies in rodent models. Feeding diets with different ratios of linoleic acid (LA) to α-linolenic acid (ALNA) during pregnancy and lactation has been shown to alter DNA methylation in the liver of the offspring at weaning. Female mice were fed diets with Control (LA:ALNA ratio (8:1)) or ALNA deficient (55:1) for 30 days before pregnancy and throughout pregnancy. Lactating dams were either maintained on the same diet they were fed in pregnancy or were switched to an ALNA supplemented (0.3:1) diet until the offspring were weaned.[28] Average methylation of the Fads2 promoter and of intron 1 did not differ between dams that were fed either the Control or ALNA deficient diet during pregnancy and lactation, but was approximately 2% higher in the liver of dams fed the ALNA supplemented diet during lactation irrespective of the diet fed during pregnancy. The ALNA supplemented diet was also associated with higher DHA concentration in maternal liver and serum. Similarly, the average methylation of the Fads2 promoter was higher in offspring of dams fed the ALNA supplemented diet than those fed the control or deficient diets throughout pregnancy and lactation. However, methylation of intron 1 was not altered significantly. However, there was no difference in DHA concentration in the brains of the offspring between the maternal dietary groups. These findings suggest that differences in capacity of the dams to synthesize DHA from ALNA reflect small variations in the epigenetic regulation of Fads2. However, although the magnitude of differences in DHA synthesis was associated with higher status, this did not appear to affect DHA accumulation in the offspring. However, ALA supplementation during lactation increased neurogenesis in the dentate gyrus of the offspring which was prevented in offspring of dams fed the ALNA deficient diet during pregnancy.[29]

Two reports have tested the effect of the amount and type of maternal dietary on the epigenetic regulation of polyunsaturated fatty acid biosynthesis in rats. Female rats were fed diets containing either 3.5%, 7% or 21% fat (w/w) enriched in either saturated and monounsaturated fatty acids or in EPA+DHA.[30] The offspring were weaned on to a diet containing 4% (w/w) soybean oil. The proportions of arachidonic acid (AA) and DHA in liver and plasma phospholipids of the adult offspring were inversely associated with the amount of fat in the maternal diet irrespective of the type of fat consumed. Fads2 mRNA expression in the offspring liver was also inversely associated with maternal dietary fat intake such that methylation of some CpG loci in the Fads2 promoter was 20% higher in the offspring of dams fed 21% fat compared to those fed 3.5% fat. The second study also tested the effects of feeding dams with diets containing 7% and 21% (w/w) safflower oil, enriched in linoleic acid, or hydrogenated soybean oil, enriched in *trans* fatty acids.[31] The findings showed that proportion of AA, but not DHA, was reduced in the aorta of the offspring of dams fed 21% fat compared to those of dams fed 7% fat irrespective of the fatty acid content of the diet. This was associated with a reciprocal change in the mRNA expression of Fads1 and Fads2. There were no significant differences between dietary groups in the DNA methylation of the Fads1 promoter. However, specific CpG loci were hypermethylated in the Fads2 promoter. Mutation of one Fads2 CpG locus that was hypermethylated in offspring aorta of dams fed 21% fat and which was located within an estrogen receptor response element decreased transcriptional activity indicating that at least some hypermethylated loci were involved in regulating Fads2 transcription. Together, these findings suggest that the amount of maternal dietary fat appears to exert a greater influence on the DNA methylome of the offspring than the fatty acid composition of the diet.

Induction of hepatosteatosis by feeding mice a high fat, Westernized diet during pregnancy and after weaning was associated with increased methylation of specific CpG loci in key metabolic control genes Insig2, fatty acid synthase, PPARα and lipoprotein lipase in the liver compared to mice exposed to a low fat diet before and after weaning, or those switched between low fat and Westernized diets.[32]

Feeding pregnant rats with a high fat diet induced lower circulating Pomc and higher leptin concentration than those fed a control diet which was associated with increased methylation of the Pomc promoter in the hypothalamus in juvenile offspring fed a standard low fat diet which was increased further in adult offspring when challenged with a high fat diet.[33] These findings provide evidence that altered DNA methylation induced by altered nutrition during development may represent a mechanism for a physiological vulnerability is exacerbated by a further dietary challenge.

One study has investigated the effect of a maternal high fat diet on the expression and epigenetic regulation by DNA methylation of placental genes in mice. Female mice were fed either a control (10% energy from fat) or high fat (60% energy from fat) with undisclosed fatty acid composition for 15 days prior to mating and then during pregnancy until they were killed at day 15.5. The high fat diet induced altered methylation of specific CpG loci within the insulin-like growth factor receptor-2 differentially methylated region that differed between sexes.[34]

Feeding female mice with a diet containing either 12% energy or 62% energy derived from fat (fatty acid composition not disclosed) for 28 days before mating and throughout pregnancy and lactation induced in the adult offspring, which were fed the 12% fat diet, hypertension, hypertriglycerideamia, hyperleptinaemia, insulin resistance and lower adiponectin concentration.[35] Adiponectin mRNA expression was lower and leptin expression was higher in white adipose tissue in the offspring of dams fed the high fat diet than of those fed the control fat diet and was positively associated with the concentrations of these adipokines in blood. The adiponectin gene showed decreased levels of the transcriptionally permissive histone H3K9 acetylation was lower and the repressive histone mark dimethyl H3K9, but not methyl H4K20, was higher in the offspring of dams fed the high fat diet compared to those of dams fed the control diet. In contrast, H3K9 acetylation and dimethyl H3K9 were unchanged at the leptin promoter, while the permissive/repressive histone mark monomethyl H4K20[36,37] was increased in the offspring of dams fed the high fat diet compared to controls. This study provides the first evidence that maternal fat intake can alter transcription in a gene-specific manner by inducing persistent alterations in histone modifications. Feeding rats with a high fat diet (45% energy) compared to a control diet containing 16% energy from fat throughout pregnancy induced increased mRNA expression in neonatal liver of paraoxinase (PON)-1 and superoxide dismutase (SOD)-2 in male and female offspring, and PON3 in male offspring. This was accompanied by increased histone H4 acetylation and dimethyl H3K4, and lower trimethyl H3K9 at the PON1 promoter in male and female offspring.

mRNA expression of Npas2, a non-human primate paralog of Clock, was increased in the liver of fetal macaques of mothers fed a diet containing 32% energy from fat compared to those of mothers fed 14% energy from fat.[38] Although DNA methylation of the Npas2 promoter was low irrespective of maternal diet, the high fat diet was associated with increased H3K14 acetylation at the Npas2 promoter. Thus, deregulation of the circadian control of metabolism in liver by maternal dietary fat may occur via histone modifications alone.

Adult Animal Models

There are emerging findings that show that, at least in part, the epigenome retains plasticity throughout that life course and can be modified by dietary inputs including lipids. Adult male mice fed diets containing 35% (w/w) fat for 18 weeks had lower stearoyl-CoA-desaturase (Scd)-1 expression in their liver which was associated with increased methylation of CpG loci at -838 and -833 base pairs relative to the transcription start site compared to those fed a diet containing 4% (w/w) fat.[39] Methylation of these CpG loci was associated positively with body leanness and fat mass, plasma leptin and insulin concentrations, but associated negatively with circulating ghrelin levels. Furthermore, feeding a high fat diet containing 60% energy from fat to transgenic adult mice carrying human APOE2 induced hepatosteatosis and altered expression of 70% of the transcriptome compared to mice fed a control diet (12% energy from fat).[40] This was associated with increased levels of tri-methyl H3K9 and tri-methyl H3K4 in genes in the PPARα network which implies that down-regulation of fatty acid β-oxidation may be an important mechanism in hepatic lipid accumulation in this model. Unfortunately, these studies did not test whether such epigenetic changes persisted beyond the period of feeding the high fat diet.

Feeding adult, non-pregnant female rats a fish oil-enriched diet for nine weeks induced reduced Fads2 mRNA expression, lower concentration of AA in hepatic phospholipids and increased methylation of specific CpG loci in the Fads2 promoter compared to those fed soy bean oil.[30] However, these changes in the epigenetic regulation of Fads2 were lost when the diet was switched from fish oil to soybean oil for a further four weeks. These findings imply that although dietary lipids can alter epigenetic processes, the epigenome may revert to the state before the dietary intervention. If so, then these data mark an important distinction between nutritional inputs during development and those during adulthood; the effects of nutritional inputs during development appear to be persistent while those during adulthood may be transient. This apparent difference has implications for strategies to modify in adulthood epigenetic marks that are associated with disease traits.

Fatty Acid Intake during Pregnancy and the Epigenome of the Offspring in Humans

Lee *et al.* investigated the effect of supplementation of the diet of pregnant women (18–22 weeks gestation) with either 400mg DHA per day or olive oil on the DNA methylation of genes involved in immune function in umbilical cord blood.[41] There were no significant differences in the average methylation in any of the genes measured between women who took DHA and the placebo group, even when the groups were sub-divided for smoking and the findings were adjusted for sex, duration of gestation, BMI and batch of analysis. However, average methylation of long interspersed repetitive sequence (LINE)-1 showed approximately 1% higher methylation in cord blood from pregnancies in which women smoked in the DHA group compared to placebo group, although this difference became non-significant when the data were adjusted for sex, duration of gestation, BMI and batch of analysis. Some concerns have been raised about the design of this study.[42] The methylation status of individual CpG loci in the insulin-like growth factor receptor-2 and H19 genes were analyzed in samples from the same study.[43] One CpG locus in the IGF2 promoter 3 had 0.9% higher methylation in the DHA supplemented group compared to the olive oil supplemented group and there was no overall difference in DNA methylation between groups. Furthermore, there was no significant difference in DNA methylation between groups in the IGF2 and H19 differentially methylated regions.

Amarasekera *et al.* investigated the effect of supplementing the diet of pregnant women with a history of atopy with either 1g EPA plus 2.1g DHA or an undisclosed placebo from 20 weeks gestation until delivery on the DNA methylome of CD4+ T cells in umbilical cord blood using the Ilumina 450k array.[44] There were no significant differences between groups in any of the CpG loci or regions covered by the array and the effect sizes displayed consistently less than 5% difference.

Overall, there appears to be no evidence to support an effect of maternal supplementation with moderate or high amounts of n-3 polyunsaturated fatty acids on the DNA methylation status of genes in umbilical blood cells.

The Effects of Fatty Acid Intake on the Epigenome in Adult Humans

Recent studies have reported the effects of increasing dietary fat intake on epigenetic marks in adult humans. In a dietary crossover study, for a five-day period, young men consumed either a diet in which 60% of energy was derived from fat (one third each of monounsaturated fatty acids, polyunsaturated fatty acids and saturated fatty acids) or one in which 35% of energy was fat derived. The former diet induced altered methylation of 7,909 CpG loci in 6,508 genes in skeletal muscle compared to the latter.[45,46] The median difference of loci that with increased methylation was 2–3% (1,979 loci) and only 8 loci differed by more than 10%. CpG loci that had lower methylation after the intervention had a median difference of 4–5% (214 loci). Hypermethylated loci were associated with pathways related to cancer, reproductive system disease and gastro-intestinal disease, while hypomethylated loci associated with inflammatory disease, inflammatory response and ophthalmic disease. These changes in DNA methylation were not strongly associated with mRNA expression. Switching from the high fat to the lower fat diet reversed the methylation changes and for 5% of the altered genes, the level of methylation was significantly altered in the opposite direction to that induced by the high fat diet. Gillberg et al. have shown using a similar study design that higher fat intake induced a mean increase of approximately 2% in the methylation level of specific CpG loci in the PPAR-γ-co-activator-1α promoter in adipose in men aged 23 to 27 years who were born at or below the 10th birth weight percentile, but there was no significant change in those born between the 50th and 90th birth weight percentile.[47]

Supplementing the diet of renal patients with 4 g daily of EPA plus docosapenaenoic acid plus DHA ethyl esters or olive oil for eight weeks altered the methylation status of individual CpG loci in the 5′ regulatory regions of genes involved in polyunsaturated fatty acid biosynthesis.[48] Olive oil and n-3 polyunsaturated fatty acids induced differential changes (≥10% difference in methylation) in specific CpG loci in Fads2 and ELOVL-5, which encodes fatty acid elongase-5 contingent on sex, but not Fads1 or ELOVL2. The methylation status of the altered CpGs in Fads2 and ELOVL5 was associated negatively with the level of their transcripts. The effects of supplementation with n-3 polyunsaturated fatty acids were replicated in healthy adults (aged 23 to 30 years). These findings suggest that even modest fatty acid supplementation can altered methylation of specific CpG loci in adult humans. Whether these effects persisted beyond the period of supplementation was not tested.

Mechanisms

The mechanism by which fatty acids modify the epigenome is a matter for debate and it is possible that more than one mechanism may operate simultaneously. There is substantial evidence that short chain fatty acids can inhibit histone deactylase activity (butyric acid > valproic acid > propionic acid > caproic acid > acetic acid).[49] However, since these are derived primarily from fermentation of dietary fiber by colonic bacteria and synthesis of short chain fatty acids from dietary fat is probably only an important source in fat malabsorption, inhibition of HDACs by short chain fatty acids might be of lesser importance in healthy individuals. Furthermore, this mechanism cannot explain the effect of treatment of cultured cells with fatty acids on epigenetic processes. Variation in energy intake leading to changes in cellular NAD+/NADH which may, in turn, alter histone acetylation by modulating the activities of the histone deaceylases sirtuins.[50] However, while such effects may be involved in epigenetic changes induced by high fat diets they appear less likely as an explanation for the impact of modest changes in fatty acids on epigenetic marks in cultured cells.

Feeding pregnant and lactating mice a high fat diet induced lower expression of expression of a number of non-coding RNAs in the adult offspring.[51] Non-coding RNAs have been shown to modify DNA methylation by changing the activity of DNA methyltransferase 3A and 3B.[52] Non-coding RNAs have also been shown to alter chromatin structure by changing methyl CpG binding protein-2 activity[53] and by altering the expression of the histone methyltransferase Enhancer of Zeste homolog-2.[54] Thus the observation that exposure to a high fat diet during development altered the expression of specific non-coding RNA species suggests a putative mechanism by which fatty acid intake may modify specific epigenetic regulatory processes. It remains to be investigated whether individual fatty acids can induce differential changes in the activities of individual non-coding RNAs.

The level of DNA methylation has been suggested to be the product of the equilibrium between methylation and demethylation reactions.[2] Although the methylation of CpG dinucleotides by DNA methyltransferases is well-understood, less is known about the mechanism by which methyl groups are removed from 5-methylcytosine and a number of candidate mechanisms have been proposed.[55] There has been substantial recent interest in the potential role of ten eleven translocation proteins (TET) as a system for DNA demethylation. TET 1, 2, and 3 remove methyl groups from methylcytosine by sequential oxidation to form carboxylcytosine, followed by base excision of formylcytosine or carboxylcytosine by thymine-DNA glycosylase.[55] TET activities are positively regulated by α-ketoglutaric acid[56] and hence increased dietary fat may lead to up-regulation of DNA demethylation by increasing flux of acetyl-CoA derived from fatty acid β-oxidation to Kreb's cycle, leading to a rise in α-ketoglutaric acid concentration. Furthermore, TET1 and TET3, but not TET2, encode an N-terminal domain zinc finger cysteine-X-X-cysteine sequence which is characteristic of proteins that interact with chromatin.[57] The interaction of TET2 with DNA appears to be via IDAX which contains a zinc finger cysteine-X-X-cysteine sequence.[58] However, unlike other chromatin binding protein, TET1 and 2, and IDAX recognizes both methylated and unmethylated CpG loci.[55,59-63] Thus modulation of the activities of DNA methyltransferases and TET proteins provides a candidate mechanism by which the methylation of specific CpG loci could be reduced in a targeted manner by fatty acids either in the diet on in cell culture. However, this mechanism does not explain increased methylation induced by fatty acids, although it is possible that down-regulation of the demethylation pathway could result in increased methylation at specific CpG loci. One mechanism by which DNA demethylation could be targeted to specific CpG dinucleotides is via poly(ADP-ribosyl)ation (PARylation) of transcription factors. PARylation of PPARΥ has been shown to recruit TET activity and induce demethylation of CpG loci that are proximal to the PPARΥ response element.[64]

Conclusions and Perspectives

There is increasing evidence that dietary fatty acids can modify DNA methylation, histone modifications and non-coding RNAs.

However, there are several important limitations to progress in this area. A number of studies have shown relatively small changes, particularly in DNA methylation, and it remains to be determined whether effects are sufficient to alter transcription. Furthermore, there is a general lack of experimental evidence to show whether differential changes in epigenetic processes — including those induced by dietary fatty acids — alter gene function and physiological outcomes change gene function. However, these concerns are likely to be addressed as new technologies emerge and consequently this field of research has potential to provide novel insights into the effects of fatty acids on development, metabolism and risk of disease which, in turn, have implications for nutritional recommendations to individuals and populations.

References

1. A Bird. (2002) DNA methylation patterns and epigenetic memory. *Genes Devel* **16**:6–21.
2. M Szyf. (2007) The dynamic epigenome and its implications in toxicology. *Toxicol Sci* **100**:7–23.
3. GC Burdge, SP Hoile, KA Lillycrop. (2012) Epigenetics: are there implications for personalised nutrition? *Curr Opinn Clin Nutr Met Care* **15**:442–447.
4. GC Burdge, KA Lillycrop. (2010) Nutrition, epigenetics, and developmental plasticity: implications for understanding human disease. *Annu Rev Nutr* **30**:315–339.
5. A Santillo, M Albenzio, M Quinto *et al.* (2009) Probiotic in lamb rennet paste enhances rennet lipolytic activity, and conjugated linoleic acid and linoleic acid content in Pecorino cheese. *J Dairy Sci* **92**: 1330–1337.
6. SJ Miller. (2004) Cellular and physiological effects of short-chain fatty acids. *Mini Rev Med Chem* **4**:839–845.
7. K Steliou, MS Boosalis, SP Perrine *et al.* (2012) Butyrate histone deacetylase inhibitors. *Biores Open Access* **1**:192–198.
8. HJ Kim, P Leeds, DM Chuang. (2009) The HDAC inhibitor, sodium butyrate, stimulates neurogenesis in the ischemic brain. *J Neurochem* **110**:1226–1240.

9. ME McGee-Lawrence, JJ Westendorf. (2011) Histone deacetylases in skeletal development and bone mass maintenance. *Gene* **474**:1–11.
10. S Lee, JR Park, MS Seo *et al.* (2009) Histone deacetylase inhibitors decrease proliferation potential and multilineage differentiation capability of human mesenchymal stem cells. *Cell Prolif* **42**:711–720.
11. P Mali, BK Chou, J Yen *et al.* (2010) Butyrate greatly enhances derivation of human induced pluripotent stem cells by promoting epigenetic remodeling and the expression of pluripotency-associated genes. *Stem Cells* **28**:713–720.
12. S Wu, RW Li, W Li, CJ Li. (2012) Transcriptome characterization by RNA-seq unravels the mechanisms of butyrate-induced epigenomic regulation in bovine cells. *PLoS One* **7**:e36940.
13. FO Andrade, MK Nagamine, AD Conti *et al.* (2012) Efficacy of the dietary histone deacetylase inhibitor butyrate alone or in combination with vitamin A against proliferation of MCF-7 human breast cancer cells. *Bra J Med Biol Res* **45**:841–850.
14. CC Spurling, JA Suhl, N Boucher *et al.* (2008) The short chain fatty acid butyrate induces promoter demethylation and reactivation of RARbeta2 in colon cancer cells. *Nutr Cancer* **60**:692–702.
15. B Kantor, H Ma, J Webster-Cyriaque *et al.* (2009) Epigenetic activation of unintegrated HIV-1 genomes by gut-associated short chain fatty acids and its implications for HIV infection. *Proc Natl Acad Sci USA* **106**:18786–18791.
16. W Almutawaa, NH Kang, Y Pan, LP Niles. (2014) Induction of neurotrophic and differentiation factors in neural stem cells by valproic acid. *Basic Clin Pharmacol Toxicol* **115**:216–221.
17. E Hall, P Volkov, T Dayeh *et al.* (2014) Effects of palmitate on genome-wide mRNA expression and DNA methylation patterns in human pancreatic islets. *BMC* **Med 12**:103.
18. JM Maples, JJ Brault, BM Shewchuk *et al.* (2015) Lipid exposure elicits differential responses in gene expression and DNA methylation in primary human skeletal muscle cells from severely obese women. *Physiol Genomics* **47**:139–146.
19. N Sadli, ML Ackland, D De Mel *et al.* (2012) Effects of zinc and DHA on the epigenetic regulation of human neuronal cells. *Cellular Physiol Biochem* **29**:87–98.

20. V Ceccarelli, S Racanicchi, MP Martelli et al. (2011) Eicosapentaenoic acid demethylates a single CpG that mediates expression of tumor suppressor CCAAT/enhancer-binding protein delta in U937 leukemia cells. *J Biol Chem* **286**:27092–27102.
21. V Ceccarelli, G Nocentini, M Billi et al. (2014) Eicosapentaenoic acid activates RAS/ERK/C/EBPbeta pathway through H-Ras intron 1 CpG island demethylation in U937 leukemia cells. *PLoS One* **9**:e85025.
22. AJ Sinclair, MA Crawford. (1972) The accumulation of arachidonate and docosahexaenoate in the developing rat brain. *J Neurochem* **19**:1753–1758.
23. EM Tanner. (1989) *Foetus into man*. Ware: Castlemead Publications.
24. SM Innis. (1991) Essential fatty acids in growth and development. *Prog Lipid Res* **30**:39–103.
25. E Herrera, E Amusquivar, I Lopez-Soldado, H Ortega. (2006) Maternal lipid metabolism and placental lipid transfer. *Horm Res* **65 Suppl 3**:59–64.
26. GC Burdge, AN Hunt, AD Postle. (1994) Mechanisms of hepatic phosphatidylcholine synthesis in adult rat: effects of pregnancy. *Biochem J* **303**:941–947.
27. AD Postle, MD Al, GC Burdge, G Hornstra. (1995) The composition of individual molecular species of plasma phosphatidylcholine in human pregnancy. *Early Hum Dev* **43**:47–58.
28. MD Niculescu, DS Lupu, CN Craciunescu. (2013) Perinatal manipulation of alpha-linolenic acid intake induces epigenetic changes in maternal and offspring livers. *FASEB J* **27**:350–358.
29. MD Niculescu, DS Lupu, CN Craciunescu. (2011) Maternal alpha-linolenic acid availability during gestation and lactation alters the postnatal hippocampal development in the mouse offspring. *J Devel Neurosci* **29**:795–802.
30. SP Hoile, NA Irvine, CJ Kelsall et al. (2012) Maternal fat intake in rats alters 20:4n-6 and 22:6n-3 status and the epigenetic regulation of Fads2 in offspring liver. *J Nutr Biochem* **24**:1213–1220.
31. CJ Kelsall, SP Hoile, NA Irvine et al. (2012) Vascular dysfunction induced in offspring by maternal dietary fat involves altered arterial polyunsaturated fatty acid biosynthesis. *PLoS One* **7**:e34492.

32. MG Pruis, A Lendvai, VW Bloks et al. (2014) Maternal western diet primes non-alcoholic fatty liver disease in adult mouse offspring. *Acta Physiol* **210**:215–227.
33. A Marco, T Kisliouk, T Tabachnik et al. (2014) Overweight and CpG methylation of the Pomc promoter in offspring of high-fat-diet-fed dams are not "reprogrammed" by regular chow diet in rats. *FASEB J* **28**:4148–4157.
34. C Gallou-Kabani, A Gabory, J Tost et al. (2010) Sex- and diet-specific changes of imprinted gene expression and DNA methylation in mouse placenta under a high-fat diet. *PLoS One* **5**:e14398.
35. H Masuyama, Y Hiramatsu. (2012) Effects of a high-fat diet exposure in utero on the metabolic syndrome-like phenomenon in mouse offspring through epigenetic changes in adipocytokine gene expression. *Endocrinol* **153**:2823–2830.
36. D Karachentsev, K Sarma, D Reinberg, R Steward. (2005) PR-Set7-dependent methylation of histone H4 Lys 20 functions in repression of gene expression and is essential for mitosis. *Genes Devel* **19**:431–435.
37. H Talasz, HH Lindner, B Sarg, W Helliger. (2005) Histone H4-lysine 20 monomethylation is increased in promoter and coding regions of active genes and correlates with hyperacetylation. *J Biol Chem* **280**:38814–38822.
38. M Suter, P Bocock, L Showalter et al. (2011) Epigenomics: maternal high-fat diet exposure in utero disrupts peripheral circadian gene expression in nonhuman primates. *FASEB J* **25**:714–726.
39. RW Schwenk, W Jonas, SB Ernst et al. (2013) Diet-dependent alterations of hepatic Scd1 expression are accompanied by differences in promoter methylation. *Horm Metabolic Res* **45**:786–794.
40. HJ Jun, J Kim, MH Hoang, SJ Lee. (2012) Hepatic lipid accumulation alters global histone h3 lysine 9 and 4 trimethylation in the peroxisome proliferator-activated receptor alpha network. *PLoS One* **7**:e44345.
41. HS Lee, A Barraza-Villarreal, H Hernandez-Vargas et al. (2013) Modulation of DNA methylation states and infant immune system by dietary supplementation with omega-3 PUFA during pregnancy in an intervention study. *Am J Clin Nutr* **98**:480–487.

42. GC Burdge, PC Calder. (2013) Does early n-3 fatty acid exposure alter DNA methylation in the developing human immune system? *Clin Lipidol* **8**:505–508.
43. HS Lee, A Barraza-Villarreal, C Biessy *et al.* (2014) Dietary supplementation with polyunsaturated fatty acid during pregnancy modulates DNA methylation at IGF2/H19 imprinted genes and growth of infants. *Physiol Genomics* **46**:851–857.
44. M Amarasekera, P Noakes, D Strickland *et al.* (2014) Epigenome-wide analysis of neonatal CD4(+) T-cell DNA methylation sites potentially affected by maternal fish oil supplementation. *Epigenetics* **9**:1570–1576.
45. C Brons, CB Jensen, H Storgaard *et al.* (2009) Impact of short-term high-fat feeding on glucose and insulin metabolism in young healthy men. *J Physiol* **587**:2387–2397.
46. SC Jacobsen, C Brons, J Bork-Jensen *et al.* (2012) Effects of short-term high-fat overfeeding on genome-wide DNA methylation in the skeletal muscle of healthy young men. *Diabetologia* **55**:3341–3349.
47. L Gillberg, SC Jacobsen, T Ronn *et al.* (2014) PPARGC1A DNA methylation in subcutaneous adipose tissue in low birth weight subjects — impact of 5 days of high-fat overfeeding. *Metabolism* **63**:263–271.
48. SP Hoile, R Clarke-Harris, RC Huang *et al.* (2014) Supplementation with N-3 long-chain polyunsaturated fatty acids or olive oil in men and women with renal disease induces differential changes in the DNA methylation of FADS2 and ELOVL5 in peripheral blood mononuclear cells. *PLoS One* **9**:e109896.
49. JR Davie. (2013) Inhibition of histone deacetylase activity by butyrate. *J Nutr* **133**:2485S–2493S.
50. S Imai, CM Armstrong, M Kaeberlein, L Guarente. (2000) Transcriptional silencing and longevity protein Sir2 is an NAD-dependent histone deacetylase. *Nature* **403**:795–800.
51. J Zhang, F Zhang, X Didelot *et al.* (2009) Maternal high fat diet during pregnancy and lactation alters hepatic expression of insulin like growth factor-2 and key microRNAs in the adult offspring. *BMC Genomics* **10**:478.
52. M Fabbri, R Garzon, A Cimmino *et al.* (2007) MicroRNA-29 family reverts aberrant methylation in lung cancer by targeting DNA methyltransferases 3A and 3B. *Proc Natl Acad Sci USA* **104**:15805–15810.

53. ME Klein, DT Lioy, L Ma et al. (2007) Homeostatic regulation of MeCP2 expression by a CREB-induced microRNA. *Nature Neurosci* **10**:1513–1514.
54. S Varambally, Q Cao, RS Mani et al. (2008) Genomic loss of microRNA-101 leads to overexpression of histone methyltransferase EZH2 in cancer. *Science* **322**:1695–1699.
55. H Wu, Y Zhang. (2014) Reversing DNA methylation: mechanisms, genomics, and biological functions. *Cell* **156**:45–68.
56. Kaelin WG, Jr., McKnight SL. (2013) Influence of metabolism on epigenetics and disease. *Cell* **153**:56–69.
57. HK Long, NP Blackledge, RJ Klose. (2013) ZF-CxxC domain-containing proteins, CpG islands and the chromatin connection. *Biochem Soc Trans* **41**:727–740.
58. M Ko, J An, HS Bandukwala et al. (2013) Modulation of TET2 expression and 5-methylcytosine oxidation by the CXXC domain protein IDAX. *Nature* **497**:122–126.
59. K Williams, J Christensen, MT Pedersen et al. (2011) TET1 and hydroxymethylcytosine in transcription and DNA methylation fidelity. *Nature* **473**:343–348.
60. H Wu, AC D'Alessio, S Ito et al. (2011) Dual functions of Tet1 in transcriptional regulation in mouse embryonic stem cells. *Nature* **473**:389–393.
61. Y Xu, F Wu, L Tan et al. (2011) Genome-wide regulation of 5hmC, 5mC, and gene expression by Tet1 hydroxylase in mouse embryonic stem cells. *Mol Cell* **42**:451–464.
62. Y Xu, C Xu, A Kato et al. (2012) Tet3 CXXC domain and dioxygenase activity cooperatively regulate key genes for Xenopus eye and neural development. *Cell* **151**:1200–1213.
63. H Zhang, X Zhang, E Clark et al. (2010) TET1 is a DNA-binding protein that modulates DNA methylation and gene transcription via hydroxylation of 5-methylcytosine. *Cell Res* **20**:1390–1393.
64. K Fujiki, A Shinoda, F Kano et al. (2013) PPARgamma-induced PARylation promotes local DNA demethylation by production of 5-hydroxymethylcytosine. *Nature Comm* **4**:2262.

Circadian Biology: Interaction with Metabolism and Nutrition

CHAPTER 5

Jonathan D. Johnston

Faculty of Health and Medical Sciences,
University of Surrey,
Guildford, GU2 7XH, UK

Introduction

Multiple key aspects of our physiology are subject to daily variation. In most cases, this daily variation is driven by processes called circadian rhythms. To be truly circadian, a rhythm must fulfil a number of criteria, including being endogenously generated and occurring with a periodicity of approximately 24 hours.[1]

The endogenous nature of human circadian rhythms can be demonstrated using a number of laboratory protocols, the simplest of which is called a constant routine. The fundamental principle of a constant routine is to remove any external variation, or spread it evenly across a 24-hour cycle. As such, it requires participants to remain awake in constant dim light and temperature, maintain a semi-supine posture and receive their daily food intake as identical small snacks evenly distributed over time, typically every hour.[2] Under these stringent conditions, and despite the lack of 24-hour time cues, human circadian rhythms can be readily observed in diverse measures such as cognitive performance, hormone concentration,[3] the transcriptome,[4] and the metabolome.[5,6] From a nutritional perspective, it is perhaps particularly noteworthy that healthy individuals exhibit circadian rhythms in plasma triglyceride and glucose concentration, together with insulin sensitivity.[7]

Over the past 15–20 years, remarkable progress has been made in our understanding of circadian rhythmicity in mammals. This understanding has progressed from the core molecular nature of circadian rhythms, through to the way in which these circadian mechanisms interact with major biochemical and physiological pathways throughout the body.

Molecular Basis of Circadian Rhythms

The molecular mechanisms that underpin circadian rhythms are becoming ever more complex and are reviewed in detail elsewhere, e.g. Partch et al.[8] Rather than reproduce these reviews, the following text will summarize some of the main points.

The predominant molecular model of circadian rhythms is based around a set of auto-regulatory transcriptional-translation feedback loops.[8] At the core of this model is a primary loop involving the cyclical circadian regulation of three *PERIOD* (*PER1-3*) and two *CRYPTOCHROME* (*CRY1-2*) genes and their protein products. Transcription of these genes is stimulated by a heterodimer of CLOCK (or NPAS2) and BMAL1 (or BMAL2) binding to E-box enhancer elements. The transcribed *PER* and *CRY* mRNAs are translated in the cytoplasm and form protein complexes, which translocate back into the nucleus to inhibit CLOCK-BMAL1 mediated transcription of their own genes. This molecular loop is influenced to a large degree by post-translational modifications. In this way, protein stability, accumulation and activity is tightly regulated in order to maintain a precise circadian duration to each cycle.[8]

The primary molecular loop interacts with a number of interlocking secondary loops. The best characterized of these involves the rhythmic transcription of nuclear receptor genes (*REVERBs/NR1Ds*) via CLOCK-BMAL1 mediated transcription; the resulting REVERB/NR1D proteins then regulate BMAL1 transcription and thus feedback on the core loop.[9] Other nuclear receptors (e.g. PPARα) are believed to form similar loops in specific tissues such as the liver.[10] Many of these secondary loops permit direct interactions between the circadian molecular machinery and intracellular metabolism. For example, cellular redox state is detected by heme molecules, which are endogenous ligands for REVERB/NR1D nuclear receptors.[11]

In order to drive the broad nature of physiological circadian rhythms, the molecular clock loops need to be able to interact with molecular output pathways. One mechanism that permits this is the binding of circadian proteins to regulatory elements (e.g. E-boxes) found in output genes that are not themselves part of the clock mechanism.[8] Importantly, many of these output genes encode transcription factors that regulate the expression of multiple downstream genes. Through such mechanisms, tens of thousands of genes exhibit circadian rhythmicity in the body.[4,12]

The Circadian Timing System

The first clock structures to be discovered in mammals were the suprachiasmatic nuclei (SCN), located in the anterior hypothalamus. Functional requirement of intact SCN for behavioral rhythms was first discovered using classical neuronal lesioning experiments.[13,14] Later work demonstrated that transplant of SCN into lesioned animals restores behavioral rhythms and period length is determined by period of donor animal, not the host.[15]

For many years, it was believed that SCN could be the sole circadian clock within mammals. However, this concept was challenged by physiological studies, e.g. measurement of circadian melatonin secretion by retinal cultures.[16] It is in fact now realized that functional clocks exist in nearly all mammalian tissues, and quite possibly in the majority of cells.[17] Clocks outside the SCN are termed peripheral clocks. The existence of these peripheral clocks is elegantly demonstrated by the use of transgenic animals in which a reporter gene (e.g. firefly luciferase) is driven by the regulatory elements of an endogenous clock gene.[18,19] Real-time measurement of reporter gene activity in cultured cells and tissue explants from these animals shows that circadian clocks exist throughout the body. Furthermore, similar rhythms can be observed in cultured human cells following use of lentiviral particles to introduce circadian reporter constructs.[20,21]

Together, the SCN clock and peripheral clocks form a complex circadian timing system. In order for this system to function optimally, all of the body's circadian clocks must be synchronized to one another. The maintenance of this internal synchrony requires the SCN, which is able to control the phase of peripheral clocks via multiple rhythmic output pathways.[22] These outputs include neuronal tone, hormone secretion, altered core body temperature and the regulation of behaviors such as sleep-wake and feed-fast cycles. Due to this central co-ordinating role of the SCN clock, it is often referred to as the "master" clock in mammals.[22] The importance of timed feeding will be described in detail below.

Role of Peripheral Clocks in Metabolic Physiology

Indirect evidence for the role of peripheral clocks came from a series of microarray studies that investigated the circadian transcriptome and proteome. These studies typically concluded that up to 10% of all transcripts or proteins in a given tissue exhibit circadian rhythmicity.[23] More recently, it has been estimated that nearly 50% of the entire mouse transcriptome is rhythmic in at least one tissue of the body.[12] Analysis of the reported functions of circadian molecules highlights that they commonly encode rate-limiting factors in important local physiological and metabolic pathways.[23]

The function of clocks in individual tissues can be more directly demonstrated using various experimental approaches. Analysis of cultured cells indicates that the secretion of some important metabolic hormones (e.g. insulin and leptin) may be regulated by local circadian clocks.[24,25] Detailed examination of adipose tissue *in vivo* and *in vitro* also indicates that peripheral clocks drive temporal changes in key metabolic pathways.[26]

Although cell culture provides clear evidence of function in isolated cell populations, it is not sufficient to reveal function of these clocks within the complexity of whole body physiology. In order to overcome this issue, various groups have taken advantage of updates in mouse genetics to produce animals that lack a functional clock in only one tissue or cell type. For example, tissue-specific clock disruption in liver, pancreas and skeletal muscle disrupts glucose homeostasis. The liver clock appears to regulate glucose uptake and fasting glucose release into blood[27]; the pancreas clock regulates glucose-stimulated insulin secretion[28,29]; the skeletal muscle clock regulates insulin-stimulated glucose uptake and intracellular glucose metabolism.[30] Of the other transgenic models generated, the clock disruption in white adipose tissue (WAT) is of particular relevance to nutrition and metabolism. Unlike other models, mice lacking a functional WAT clock become obese; the proposed mechanism for this phenotype involves abnormal WAT fatty acid metabolism indirectly altering circadian rhythms within hypothalamic appetite centers and thus a less pronounced feed-fast cycle.[31]

Together these studies are rapidly building up a picture of the functional importance of peripheral clocks. Further transgenic models are likely to appear very soon to elucidate clock function in other tissues relevant to nutritional metabolism, including the intestine and brain regions central to feeding behavior.

Meal Timing and Chrononutrition

Our increased understanding of the basic biology linking circadian and metabolic physiology has a number of potential practical outlets. One of the most obvious of these is the application to nutritional science. This interaction between circadian rhythms and nutrition is often referred to as "chrononutrition".

One well-studied aspect of chrononutrition is the variation of post-prandial responses across the circadian cycle. Eating during the biological evening and night results in elevated post-prandial plasma concentrations of glucose and triglyceride that take longer to return to basal levels.[7] Although the long-term consequences of these elevated post-prandial responses are not well-studied, it seems highly plausible that they could contribute to increased incidence of obesity and metabolic disorders reported in certain risk groups including shift workers and individuals with night eating syndrome.[32]

In addition to the influence of circadian rhythms on metabolism and nutritional physiology is the effect of food consumption on circadian rhythmicity. It has been recognized for many decades in model species that temporal restriction of food to a short time window each day results in an anticipatory bout of locomotor activity and physiological changes.[33,34] These anticipatory changes exhibit their own circadian properties, but persist in the absence of function SCN, at least in rodents. The existence of a food entrainable oscillator (FEO) has thus been postulated, but its anatomical location(s) remain unknown. As reviewed elsewhere,[32] neuronal lesion studies have revealed some candidate structures related to the FEO. Also of note is the observation that temporal food restriction exerts powerful synchronization effects on clock gene expression in rodent peripheral tissues.[35] The FEO may therefore represent the interaction between multiple central and peripheral sites, without possessing a single anatomical location. To date, the effects of temporal food restriction on human circadian rhythms are poorly understood.

Conclusions and Perspectives

An impressive combination of molecular and physiological studies has clearly identified functional links between circadian, metabolic and nutritional biology. Despite recent advances, there remain a number of gaps in our understanding of basic biological mechanisms. Further tissue-specific transgenic models will reveal the *in vivo* function of clocks in key tissues, such as cells along the gastrointestinal tract. It will also be important to increase understanding of biochemical pathways that link, for instance, the transcriptional-translational circadian loops to intracellular metabolic oscillators.

In parallel with further advances in basic biology must come translation into practical outputs, particularly within the emerging field of chrononutrition. There is to date a relative paucity in human studies in this field. The development of protocols to study human molecular rhythms *in vitro*[20,21] and also *in vivo*[4–6,36] now allows for more mechanistic approaches. Ultimately it is hoped and expected that this fruitful area of research will lead to beneficial changes in public and occupational health policy, together with the development of nutritional products that are optimized for consumption at specific times and/or by particular population groups.

References

1. CS Pittendrigh. (1993) Temporal organization: reflections of a Darwinian clock-watcher. *Annu Rev Physiol* **55**:16–54.
2. JF Duffy, DJ Dijk. (2002) Getting through to circadian oscillators: why use constant routines? *J Biol Rhythms* **17**:4–13.
3. DJ Skene, J Arendt. (2006) Human circadian rhythms: physiological and therapeutic relevance of light and melatonin. *Ann Clin Biochem* **43**:344–353.
4. EE Laing, JD Johnston, CS Moller-Levet, G Bucca *et al.* (2015) Exploiting human and mouse transcriptomic data: identification of circadian genes and pathways influencing health. *Bioessays* **37**: 544–556.
5. R Dallmann, AU Viola, L Tarokh, C Cajochen *et al.* (2012) The human circadian metabolome. *Proc Natl Acad Sci USA* **109**:2625–2629.
6. SK Davies, JE Ang, VL Revell, B Holmes *et al.* (2014) Effect of sleep deprivation on the human metabolome. *Proc Natl Acad Sci USA* **111**:10761–10766.
7. L Morgan, S Hampton, M Gibbs, J Arendt. (2003) Circadian aspects of postprandial metabolism. *Chronobiol Int* **20**:795–808.
8. CL Partch, CB Green, JS Takahashi. (2014) Molecular architecture of the mammalian circadian clock. *Trends Cell Biol* **24**:90–99.
9. N Preitner, F Damiola, L Lopez-Molina, J Zakany *et al.* (2002) The orphan nuclear receptor REV-ERBalpha controls circadian transcription within the positive limb of the mammalian circadian oscillator. *Cell* **110**:251–260.
10. L Canaple, J Rambaud, O Dkhissi-Benyahya, B Rayet *et al.* (2006) Reciprocal regulation of brain and muscle Arnt-like protein 1 and peroxisome proliferator-activated receptor alpha defines a novel positive feedback loop in the rodent liver circadian clock. *Mol Endocrinol* **20**:1715–1727.
11. L Yin, N Wu, JC Curtin, M Qatanani *et al.* (2007) Rev-erbalpha, a heme sensor that coordinates metabolic and circadian pathways. *Science* **318**:1786–1789.
12. R Zhang, NF Lahens, HI Ballance, ME Hughes *et al.* (2014) A circadian gene expression atlas in mammals: implications for biology and medicine. *Proc Natl Acad Sci USA* **111**:16219–16224.

13. FK Stephan, I Zucker. (1972) Circadian rhythms in drinking behavior and locomotor activity of rats are eliminated by hypothalamic lesions. *Proc Natl Acad Sci USA* **69**:1583–1586.
14. RY Moore, VB Eichler. (1972) Loss of a circadian adrenal corticosterone rhythm following suprachiasmatic lesions in the rat. *Brain Res* **42**:201–206.
15. MR Ralph, RG Foster, FC Davis, M Menaker. (1990) Transplanted suprachiasmatic nucleus determines circadian period. *Science* **247**:975–978.
16. G Tosini, M Menaker. (1996) Circadian rhythms in cultured mammalian retina. *Science* **272**:419–421.
17. C Dibner, U Schibler. (2015) Circadian timing of metabolism in animal models and humans. *J Intern Med* **277**:513–527.
18. S Yamazaki, R Numano, M Abe, A Hida *et al.* (2000) Resetting central and peripheral circadian oscillators in transgenic rats. *Science* **288**:682–685.
19. SH Yoo, S Yamazaki, PL Lowrey, K Shimomura *et al.* (2004) PERIOD2::LUCIFERASE real-time reporting of circadian dynamics reveals persistent circadian oscillations in mouse peripheral tissues. *Proc Natl Acad Sci USA* **101**:5339–5346.
20. SA Brown, F Fleury-Olela, E Nagoshi, C Hauser *et al.* (2005) The period length of fibroblast circadian gene expression varies widely among human individuals. *PLoS Biol* **3**:e338.
21. S Hasan, N Santhi, AS Lazar, A Slak *et al.* (2012) Assessment of circadian rhythms in humans: comparison of real-time fibroblast reporter imaging with plasma melatonin. *FASEB J* **26**:2414–2423.
22. U Albrecht. (2012) Timing to perfection: the biology of central and peripheral circadian clocks. *Neuron* **74**:246–260.
23. GE Duffield. (2003) DNA microarray analyses of circadian timing: the genomic basis of biological time. *J Neuroendocrinol* **15**:991–1002.
24. E Peschke, D Peschke. (1998) Evidence for a circadian rhythm of insulin release from perifused rat pancreatic islets. *Diabetologia* **41**:1085–1092.
25. DT Otway, G Frost, JD Johnston. (2009) Circadian rhythmicity in murine pre-adipocyte and adipocyte cells. *Chronobiol Int* **26**:1340–1354.

26. A Shostak, J Meyer-Kovac, H Oster. (2013) Circadian regulation of lipid mobilization in white adipose tissues. *Diabetes* **62**:2195–2203.
27. KA Lamia, KF Storch, CJ Weitz. (2008) Physiological significance of a peripheral tissue circadian clock. *Proc Natl Acad Sci USA* **105**: 15172–15177.
28. B Marcheva, KM Ramsey, ED Buhr, Y Kobayashi *et al.* (2010) Disruption of the clock components CLOCK and BMAL1 leads to hypoinsulinaemia and diabetes. *Nature* **466**:627–631.
29. LA Sadacca, KA Lamia, AS deLemos, B Blum *et al.* (2011) An intrinsic circadian clock of the pancreas is required for normal insulin release and glucose homeostasis in mice. *Diabetologia* **54**:120–124.
30. KA Dyar, S Ciciliot, LE Wright, RS Bienso *et al.* (2014) Muscle insulin sensitivity and glucose metabolism are controlled by the intrinsic muscle clock. *Mol Metab* **3**:29–41.
31. GK Paschos, S Ibrahim, WL Song, T Kunieda *et al.* (2012) Obesity in mice with adipocyte-specific deletion of clock component Arntl. *Nat Med* **18**:1768–1777.
32. JD Johnston. (2014) Physiological responses to food intake throughout the day. *Nutr Res Rev* **27**:107–118.
33. FK Stephan. (2002) The "other" circadian system: food as a Zeitgeber. *J Biol Rhythms* **17**:284–292.
34. RE Mistlberger. (2009) Food-anticipatory circadian rhythms: concepts and methods. *Eur J Neurosci* **30**:1718–1729.
35. F Damiola, N Le Minh, N Preitner, B Kornmann *et al.* (2000) Restricted feeding uncouples circadian oscillators in peripheral tissues from the central pacemaker in the suprachiasmatic nucleus. *Genes Dev* **14**:2950–2961.
36. DT Otway, S Mantele, S Bretschneider, J Wright *et al.* (2011) Rhythmic diurnal gene expression in human adipose tissue from individuals who are lean, overweight, and type 2 diabetic. *Diabetes* **60**:1577–1581.

Nutrition, Epigenetics and Aging

CHAPTER 6

John C. Mathers and Hyang-Min Byun

Human Nutrition Research Centre,
Institute of Cellular Medicine, Newcastle University,
Biomedical Research Building, Campus for Ageing and Vitality,
Newcastle on Tyne, NE4 5PL, UK

Introduction: Biology of Aging

Aging has been defined as time-dependent functional decline[1] and, with very rare exceptions such as *Hydra*, all multicellular organisms demonstrate aging.[2] In humans, the effects of the aging process are observable as changes in physical form and in the capacity to undertake a wide range of functions. For example, changes in facial morphology and pigmentation associated with aging are easily recognized and encourage some to pursue plastic surgery to achieve a younger looking face.[3] A very recent study showed that, even at the same chronological age, individuals who were aging more rapidly looked older.[4] It seems likely that we have evolved to be able to judge how old other people appear since this will have advantages when choosing a mate and when assessing rivals. The biological basis for aging has been difficult to establish because it occurs very gradually and is so pervasive. In addition, for both conceptual and practical reasons, it has also proved difficult to separate aging from the pathogenesis of age-related diseases. However, most authorities now agree that biological aging results from the accumulation of multiple kinds of molecular damage.[1,5] This accumulation of molecular damage causes cellular damage/dysfunction which leads to the aging phenotype. In turn, the aging phenotype may precipitate the development of age-related diseases when molecular damage [of a particular kind] reaches a tipping point/threshold in a particular cell type.

Based on evidence from aging studies in a wide range of organisms, but especially in mammals, recently López-Otín and colleagues proposed nine tentative hallmarks of aging.[1] These include genomic instability, telomere attrition, epigenetic alterations and loss of proteostasis which López-Otín and colleagues considered to be the primary hallmarks and the causes of molecular/cellular damage. Deregulated nutrient sensing, mitochondrial dysfunction and cellular senescence are responses to the time-dependent accumulation of damage which, at least initially, may be compensatory or antagonistic but which eventually become deleterious in themselves.[1] Finally, stem cell exhaustion and altered intercellular communication were identified as integrative hallmarks and considered to be culprits of the aging process which are ultimately responsible for the functional decline observed in aging.[1]

Genetics of Longevity

There is little support for the idea that aging is programmed, i.e. that there is a genetically encoded system for limiting lifespan. That said, there is a wealth of data from studies of many organisms, from yeast to humans, showing that longevity has a strong genetic component.[6] Research with invertebrates and with mice has identified multiple genetic pathways which influence lifespan and which suggest longevity is promoted through the manipulation of metabolism and the resistance to, or capacity to repair, molecular damage.[6,7] In humans, the genetic contribution to longevity has been estimated to be 15–25%.[8–10] However, variants in only two genes *APOE* and *FOXO3* have been associated consistently with extended lifespan in multiple studies including a recent meta-analysis of results from genome-wide association studies (GWAS).[11] The latter analysis also found suggestive evidence for associations between variants in *CADM2* and in *GRIK2* (genes that are involved in neuron cell-cell adhesion and in regulation of glutamatergic synaptic transmission) and longevity but these will need confirmation in future studies.[11] As with other complex phenotypes such as adiposity, the apparent failure of GWAS approaches to identify common genetic variants which explain a significant proportion of the heritability of the trait has led to suggestions that epigenetic mechanisms might alter the expression of genetic variants influencing the aging process/longevity.[11]

Inter-individual Variation in Aging

Aging is a highly individual process. Even genetically identical organisms such as the small nematode worm *Caenorhabditis elegans* (*C. elegans*) reared in a constant environment show considerable inter-individual variation in lifespan.[12] Kirkwood and Finch have suggested that this is evidence for the contribution of stochastic damage to molecules and cells which supports the disposable soma theory of aging.[13] The latter is based on evolutionary physiology and argues that how organisms allocate resources between reproduction and maintenance and repair of cells and tissues determines longevity with longer-lived organisms devoting a higher proportion of resources to maintenance of the soma.[5] In pursuit of an explanation for the stochastic contribution to inter-individual differences in lifespan, Kirkwood et al. proposed a number of sources of intrinsic variation at molecular, cellular, organ/system and, for social species, community levels which might operate during development and/or during adulthood.[5]

One would anticipate that inter-individual differences in capacity to cope with molecular damage would contribute to inter-individual differences in aging. For example, because of its paramount role in determining gene expression and cell function, damage to DNA can be particularly devastating. Therefore, mammals have evolved an elaborate genomic maintenance apparatus with multiple DNA repair systems, including base excision repair (BER) and nucleotide excision repair (NER), which recognize and repair specific genomic lesions.[14] Individuals with rare inherited defects in DNA repair systems, e.g. those with xeroderma pigmentosum, Cockayne's Syndrome, the Werner Syndrome and trichothiodystrophy experience accelerated aging.[14] Whilst many of the genes encoding DNA repair systems are polymorphic, there is little evidence that non-pathological variants in DNA repair genes are associated with reduced lifespan.[15] However, at a functional level, there is evidence of considerable inter-individual variation in DNA repair capacity. For example, even among healthy young adults (aged 18–30 years), we observed 11-fold variation in NER capacity between individuals and found that NER capacity in peripheral blood mononuclear cells was inversely associated with age.[16] In addition, we observed an inverse association between adiposity and NER.[16] In the same healthy young adults, there was also inter-individual variation in BER measured using the Comet assay.[17]

Nutrition and Aging

Obesity and Dietary Energy Restriction

The repeated demonstration that dietary (energy) restriction (DR) extends lifespan across taxa from yeast, to *C. elegans* and to rodents[18] provides proof of principle that nutrition is a major modifier of the aging process. However, a meta-analysis of DR studies in rodents found that this intervention did not produce longer life in some inbred mouse strains and, indeed, that DR results in shortening of lifespan in the ILSXISS strain 114.[19] In addition, the effects of DR (in the absence of malnutrition) in primates remain uncertain. An initial study of adult-onset DR at the University of Wisconsin (UW) reported reduced disease onset and reduced mortality in DR monkeys.[20] In contrast, a parallel study at the National Institute of Aging found no significant effect of DR on lifespan, although, of course, the DR animals were leaner and in better health.[21] An extended analysis of the UW study confirmed that DR reduced both age-related and all-cause mortality significantly and the authors claimed that these findings indicated that the effects of DR on aging are conserved in primates.[22]

For both ethical and logistical reasons, a randomized controlled study of the long-term effects of DR in humans is unlikely but there is good evidence from numerous observational studies that greater adiposity (usually measured as body mass index (BMI)) is associated with higher mortality risk.[23] In addition, gaining body weight in mid-life exacerbates the adverse effects of higher BMI in young adulthood on aging in later life.[24] Maintaining a healthy weight (BMI<25) through diet and physical activity will enhance healthy aging by minimizing the molecular damage caused by obesity-associated inflammation and metabolic stress.[25,26]

Dietary patterns and aging

For several decades, residents of Japan have enjoyed longer lifespans and greater health during aging than those of most other countries. Current statistics show that for males and females combined, life expectancy at birth in Japan is now 84 years and healthy life expectancy is 75 years.[27] Such observations have provoked interest in associations between the Japanese lifestyle and healthy aging[28] with attempts to identify dietary patterns associated with longer lifespan and lower disease risk in later life.[29] In addition, there is strong evidence from geographically diverse longitudinal studies that a Mediterranean dietary pattern is associated with greater longevity and lower risk of death for several common age-related diseases.[30] Evidence of causality for the beneficial effects of a Mediterranean dietary pattern in prevention of age-related disease, in particular cardiovascular disease (CVD), has been provided by both secondary[31] and primary prevention intervention studies.[32] Recent results from a study of a large, multi-ethnic population of >200,000 Americans, suggests high scores on any of four dietary patterns (viz. the Healthy Eating Index-2010, the Alternative HEI-2010, the alternate Mediterranean diet score (aMED), and the Dietary Approaches to Stop Hypertension (DASH)) are associated with lower risk of mortality from all causes and from CVD and cancer.[33] It seems probable that many (traditional) dietary patterns, all of which are characterized by high intakes of vegetables and fruit and limited intakes of animal products (excluding fish), have similar beneficial effects on aging.[34]

Epigenetics and Aging

As noted above, epigenetic alterations have been proposed as one of the primary hallmarks of aging.[1] These alterations can be aggregated into two conceptually different groups:

(i) epigenetic changes which contribute to epigenetic diversity within and between individuals over time;
(ii) epigenetic changes which proceed in a concordant fashion with aging and which may prove useful as markers of age or of biological aging.

Inter-individual Variation in Epigenetic Marks during Aging

Epigenetic changes which contribute to epigenetic diversity within and between individuals over time have been demonstrated convincingly by studies of monozygotic (MZ) twins which show very similar epigenetic patterns in early childhood.[35] However, both histone acetylation patterns and the genomic distribution of 5-methylcytosine in DNA diverge with age.[35] It appears that at least some of these age-related, inter-twin discordances in epigenetic marks are associated with environmental exposures since twins who had spent less of their lifetime together and/or had more different medical histories showed the greatest differences in patterns of acetylation of histones H3 and H4 and in DNA methylation.[35] This study by Fraga and colleagues involved small numbers of MZ twin pairs over a relatively narrow age range and used a cross-sectional study design. A more recent study of 230 MZ twin pairs aged 18–89 years from Denmark and The Netherlands reported both cross-sectional and longitudinal data on global DNA methylation (estimated as long interspersed element 1 (LINE1) methylation) and methylation of a small panel of both imprinted and non-imprinted loci.[36] Although LINE1 methylation assesses only approximately 17% of the genome and is restricted to a particular type of transposable element, LINE1 methylation seems likely to be an acceptable surrogate for total 5-methylcytosine in DNA in many cases.[37] In the Danish-Dutch study, within-pair differences in LINE1 methylation were small but increased with age.[36] Most of this increased discordance could be attributed to environmental factors unique to each twin.[36] Such changes in global DNA methylation assayed by LINE1 methylation are difficult to interpret mechanistically and their implications for function are unclear. In contrast, inter-twin differences in methylation at specific gene loci were larger and with greater inter-individual variation over time which was estimated to be 3–16% per decade, depending on gene.[36] These findings support the idea that epigenetic changes at both imprinted and non-imprinted loci occur across the whole (adult) lifespan.[36] However their impact on gene expression was not assessed in this study and remains unclear.[36] In cross-sectional studies of unrelated individuals, we have observed increasing inter-individual variation in methylation of genes across the lifespan from birth to age 85 years.[38] Among 377 participants all aged 85 years in the Newcastle 85+ Study, quantitative analysis by pyrosequencing showed extensive and highly variable methylation of promoter-associated CpG islands with levels ranging from 4–35%, even at known tumor suppressor genes such as *TWIST2*.[38] The inter-individual differences in methylation observed in the Newcastle 85+ Study participants phenocopy multiple features of the aberrant methylation patterns typical of cancer cells and we suggested that the accumulation with time of methylation changes in promoter-associated CpG islands may contribute to the substantial increased risk of cancer with age.[38] Such methylation changes may also contribute to the development of age-related frailty.[39]

On theoretical grounds, we have argued that these time-related changes in methylation at specific loci (and, indeed, other epigenetic marks) contribute to the inter-cell heterogeneity within tissues which characterizes the aging process.[40] Such cell-to-cell variation in gene expression was described for the aging mouse heart muscle nearly 10 years ago.[41] Technological developments which allow investigation of epigenetic marks and molecules and of gene transcription and protein expression at the single-cell level[42] are likely to revolutionize this aspect of aging research (see below).

DNA Methylation "Clock"

The idea that cells/organisms contain a biological "clock" which keeps track of the individual's aging trajectory is a beguiling concept. Over the past few years, several authors have proposed DNA methylation "clocks" for assessing human aging (Table 1). Using methylation data from Illumina 27K array analysis of DNA from saliva donated by 34 male MZ twin pairs aged 21–55 years, Bocklandt and colleagues identified 88 CpG sites in or near 80 genes for which methylation correlated significantly with age, an association which was replicated in a further 31 males and 29 females aged 18–70 years.[43] Methylation of just two cytosines from these 88 loci explained 73% of the variance in age. In contrast with the MZ twin study by Fraga and colleagues, inter-twin differences

Table 1. Proposed DNA methylation "clocks" for assessing human aging.

Reference	No. of CpG sites	Tissue specificity	Analysis platform	Correlation with chronological age (Marioni et al., 2015)[44]
43	88	Saliva	Illumina 27K array	Not tested
45	71	Blood	Illumina 450K array	Good correlation (0.83). Predicted ages were slightly higher than chronological ages.
46	353	Multiple tissues	Illumina 27K and 450K arrays	Good correlation (0.75). Predicted ages were slightly lower than chronological ages for LBC1921 and LBC1936 but similar for FHS and NAS.
51	3	Blood	Illumina 27K and 450K arrays	Low correlations and large differences between predicted and chronological ages.

Note: LBC1921 and LBC1936: Lothian Birth Cohorts 1921 and 1936 respectively; FHS: Framingham Heart Study; NAS Normative Aging Study.

in methylation were small.[35] Bocklandt and colleagues[43] argued that this was because they focused on CpG sites close to functional gene transcription start sites whereas Fraga and colleagues[35] investigated random sites, most of which were located in non-functional, repeated sequences. Further, Bocklandt and colleagues suggested that critical regulatory domains within the genome remain under strict epigenetic control throughout life.[43]

More recently, three independent studies were published, each of which proposed different panels of CpG sites for which methylation might be useful in assessing human aging (Table 1). The predictive utility of these panels has been assessed by using data from four major human cohorts viz. the 1921 and 1936 Lothian Birth Cohorts (LBC) in the UK and the Framingham Heart Study (FHS) and the Normative Study of Aging (NAS) in the USA (Table 1).[44] Overall, both the Hannum et al. and Horvath panels provided good predictions of chronological age (Table 1).[45,46] Within populations there were considerable inter-individual differences between predicted and chronological age and this difference between "methylation age" and chronological age (Δ_{age}) (estimated by either the Hannum or Horvath approaches) was predictive of all-cause mortality in later life.[44] As for longevity, Δ_{age} appeared to be heritable with approximately 40% of inter-individual differences in Δ_{age} attributable to genetic factors.[44] Perhaps surprisingly, well-established life-course predictors of the aging process and of life-expectancy including APOE genotype, education, childhood IQ, social class, diabetes, high blood pressure and CVD were not associated with Δ_{age}.[44] Cross-sectional analysis of data from LBC1936 at age 70 indicated that values of Δ_{age} above chronological age were associated with poorer physical capability (hand grip strength and lung function) and reduced cognitive ability.[44] However, in longitudinal analysis at ages 73 and 76 years, Δ_{age} did not predict decline in physical or cognitive ability in LBC1936 which may be a consequence of limited power due to the relatively short follow-up.[47]

The mechanism(s) through which methylation of certain CpG sites appears to keep track of biological aging is/are unknown. Horvath suggested that DNA methylation (DNAm) age measures "the cumulative work done by a particular type of epigenetic maintenance system (EMS) which helps to maintain epigenetic stability".[46] It follows that variation in the effectiveness of the EMS (due to genetic, environmental or stochastic reasons) would be reflected in higher or lower DNAm than would be anticipated from chronological age.[46] Whilst the biological system encapsulated in the EMS proposal remains obscure, recent studies by Horvath and colleagues have shown that the accelerated aging observed in Down syndrome is matched by higher DNAm in blood and brain tissue.[48] In addition, HIV-1 infection appears to accelerate aging (measured by DNAm) by 14–15 years.[49] The hypothesis that not all tissues age at equivalent rates was supported by analysis of 30 tissues from mid-age and older adults including supercentenarians (>110 years) which showed that the cerebellum aged more slowly than other tissues.[50]

Future Perspectives

Mitochondria, Nutrition and Aging

All mammalian cells, except red blood cells, contain mitochondria with the number of mitochondria varying widely depending on cell type. Mitochondrial proteins are encoded by both the nuclear and the mitochondrial genomes. In contrast with the nuclear genome, the human mitochondrial genome is small (16.6 kb) containing just 37 genes encoding two rRNAs, 22 tRNAs and 13 polypeptides. As noted earlier, mitochondrial dysfunction is a hallmark of aging[1] and, during aging, several kinds of mutation accumulate in mitochondrial DNA (mtDNA).[52] Because of the central role played by mitochondria in several key cellular processes including ATP generation, cell signaling (especially signaling via reactive oxygen species (ROS)),[53] and regulation of apoptosis, it is not surprising that this age-associated increase in mtDNA mutation load is associated with reduced cellular function.[54] The mitochondrial free radical theory of aging proposes that, due to mitochondrial dysfunction, production of ROS increases with age causing pervasive damage to cellular macromolecules and including the mitochondrial genome. However, crude attempts to disrupt this vicious cycle by administration (usually in very large doses) of exogenous dietary antioxidants have been largely futile.[55] More progress is likely to be made based on better mechanistic understanding of the roles of ROS in normal and aberrant cell signaling and the ways in which these processes can be modulated by individual nutrients or by inflammation-related pathways, e.g. those associated with obesity.

Mitochondrial Epigenetics

Until recently, epigenetic features of mitochondria were largely neglected but this is an emerging field in which nutrition and nutritional status (in particular, adiposity) are likely to play important roles, especially during development and aging. Many aspects of mitochondrial epigenetics are concordant with nuclear epigenetics. For example, mitochondria utilize DNA methyltransferase (DNMT) enzymes and the ten-eleven translocation (TET) enzyme to generate 5-methylcytosine (5-mC) and 5-hydroxymethylcytosine (5-hmC), respectively. While predominantly restricted to CpG sites, non-CpG methylation occurs in mtDNA as it does in the nuclear genome.[56,57] Although the maintenance and function of mitochondrial DNA methylation appears to have a number of similarities with nuclear DNA methylation, there are unique aspects of mitochondrial epigenetics. Firstly, the mitochondrial genome occurs as a circular genome (reflecting its bacterial origin) without the histone complexes which are an integral component of the machinery regulating transcription in the nucleus. Therefore mtDNA methylation is the dominant epigenetic mark responsible for epigenetic regulation of mitochondrial gene expression. In addition, the mitochondrial genome is a space-efficient structure which lacks introns and "junk DNA" such as LINEs and short interspersed elements (SINEs; Alu) which comprise over 40% of the human nuclear genome.

There is interaction between the mitochondrial and nuclear epigenetic landscapes through several mechanisms.[58] As noted above, nuclear DNA encodes many mitochondrial proteins, including mitochondrial DNMT enzymes, and therefore altered expression of the nuclear-encoded methylation machinery may impact directly on mitochondrial DNA methylation. Further, many miRNAs encoded in the nuclear genome are transported to mitochondria, although their functions and effect on mitochondrial gene expression are poorly understood.[59] Multiple copies of the mitochondrial genome may be present in a given organelle, and mitochondrial DNA copy number is associated with methylation status of the nuclear-encoded DNA polymerase gamma A gene.[60] In addition, mitochondrial haplogroups, which are categorized by different mitochondrial DNA sequences (and which can be grouped by migration of human populations across the planet), show differences in global nuclear DNA methylation.[61] Indeed, mitochondrial miRNA can transfer to the nucleus and influence nuclear epigenetics.[62] Mitochondrial trafficking in neurons[63] may represent a means by which "orphan" mitochondria can influence epigenetic regulation of the nuclear genome in the adopted neuron. However, most mitochondrial epigenetic factors, notably cytosine methylation and hydroxymethylation of the mitochondrial genome and miRNAs expression from mitochondrially encoded genes, exert their influence predominantly within the mitochondria.[62]

Nutrition and Mitochondrial Epigenetics

Altered levels of mtDNA methylation are associated with altered mitochondrial gene expression,[64] and methylation patterns can be modulated by environmental factors, such as endocrine disruptors,[65] particulate forms of air pollution[66,67] and diet. Aberrant mitochondrial epigenetics has been reported in several tissues, including placenta, brain, blood and liver, with implications for multiple non-communicable diseases including cancers.[47] Furthermore, the therapeutic potential in modulating mitochondrial epigenetics has been demonstrated with valproic acid.[68] Because of its clinical application in treating mood disorders such as bipolar disease, the effects of valproic acid on the brain have been studied widely. Valproate inhibits histone deacetylases through specific degradation of these enzymes[69] and disrupts DNA methylation.[70] To the best of our knowledge, there have been no published papers on the effects of nutrients on mitochondrial epigenetics. Here, we will discuss possible mechanisms by which nutrition may influence mitochondrial epigenetics and how this might influence the aging process.

Dietary methyl donors

Dietary methyl donors can be transported to the mitochondria via carriers or transporters and can be catabolized in mitochondria.[71,72] Choline is a major source of methyl groups through its metabolite trimethylglycine (betaine) which contributes to the S-adenosyl methionine (SAM) synthesis pathways.[73] Mitochondrial choline is metabolized to betaine by the enzymes choline dehydrogenase and betaine aldehyde dehydrogenase, which are co-localized in the mitochondria.[74] A methyl group from betaine is transferred to homocysteine to generate dimethylglycine (DMG) and methionine,[74] and therefore betaine is important in generation and maintenance of methionine and SAM concentrations.[73] Folate is transported into the mitochondrion via the mitochondrial folate transporter SLC25A32 and these organelles contain >40% of cellular folate.[75,76] Dietary methyl donors contribute to the folate and methionine cycles,[77] producing SAM for methylation events in mitochondria. Whilst inadequate methyl donor status leads to defects in mitochondrial function,[78,79] to date there have been no reports of the effects of methyl donor supply on mitochondrial DNA methylation.

Dietary DNMT enzyme inhibitors

The DNMT family of enzymes catalyzes the transfer of methyl groups to cytosine residues to generate 5-methylcytosine using SAM as the methyl donor. There are several DNMTs, each with unique roles, but with some overlapping specificity.[80,81] DNMT enzyme inhibitors, such as azacytidine or decitidine which is a chemical analogue of cytidine, are used for treatment of several forms of leukemia.[82,83]

Compounds in foods can also inhibit DNMT enzyme activity. Probably the most studied is epigallocatechin-3-gallate (EGCG), the predominant polyphenolic compound in green tea, which reverses the hypermethylation of tumor suppressor genes in cancer cell lines leading to re-expression of the corresponding genes.[84] In pioneering work, Fang and colleagues used molecular modelling to demonstrate the intimate interactions between EGCG and DNMT through which the polyphenol inhibits DNMT activity by blocking the active site of the enzyme.[85] Several other dietary compounds also inhibit DNMT activity, thereby reducing global DNA methylation levels (Table 2).

Mitochondrial DNMTs are encoded by nuclear DNA and the mature proteins are transported to mitochondria. A number of different mitochondrial DNMTs have been characterized,[92,93] but the effects of dietary factors on mtDNMT activity is not known. However, we predict that such effects will occur because the nuclear and mitochondrial DNMTs share near identical structures; the only difference is the addition of an N-terminal sequence to mitochondrial DNMTs to direct their transport to the mitochondria.[92] In addition, effects of dietary compounds on expression of nuclear-encoded *mtDNMT* have not been described but are likely to occur in a manner similar to that observed with nuclear-encoded *DNMTs*.

Dietary antioxidants

As noted above, mitochondria are an important source of ROS with the potential to damage macromolecules in the cytosol and nucleus as well as the mitochondrion. Abnormal epigenetics patterns, and especially aberrant DNA methylation, are associated with increased oxidative stress.[94] Since the mitochondrial genome is in close proximity to the source of ROS generation, we predict that it will be more susceptible to altered mtDNA methylation as a consequence of oxidative damage to mtDNA *per se* or to mtDNMTs. Many dietary antioxidants, including vitamins C and E, carotenoids, polyphenols and trace metals (e.g. selenium) have the potential to reduce oxidative

Table 2. Food-derived compounds which affect DNA methylation.

Compound	Food source(s)	Active dose range	Effect on DNA methylation	Reference
Catechin and epicatechin	tea	0.2 ~ 8.1 μM	A dose-dependent inhibition of DNA methylation in gene promoter	86
Quercetin, fisetin, and myricetin	fruits, vegetables, leaves and grains	0.7 ~ 1.6 μM	A dose-dependent inhibition of DNA methylation in gene promoter	86
Caffeic acid	coffee and argan oil	3.0 ~ 13.5 μM	A dose-dependent inhibition of DNA methylation in gene promoter	87
Chlorogenic acid	potatoes and green coffee beans	0.75 ~ 2.8 μM	A dose-dependent inhibition of DNA methylation in gene promoter	87
Curcumin	Indian curry and American mustard	7.5 ~ 10 μM	Genome-wide DNA methylation changes	88
Parthenolide	flowers and fruit	5 ~ 30 μM	Global and gene promoter DNA hypomethylation	89
Mahanine	curry tree	10 ~ 20 μM	DNA methylation in gene promoter	90
Genistein	soybean and a phytoestrogen	60 ~ 100 μM	Global DNA hypomethylation	91

damage or decrease the generation of ROS but, as noted above, in most cases clinical trials with such supplements have not produced the anticipated health benefits.[55] However, our recent systematic review and meta-analysis of the effects of supplementation with vitamin C and vitamin E suggests that some population sub-groups (particularly those with low status for the nutrient) may benefit from such supplementation.[95] Antioxidants can modulate epigenetic patterns, both global and gene-specific DNA methylation, but the underlying mechanisms remain to be revealed.[96–98]

Mitochondrial epigenetics and aging

There is a significant body of knowledge regarding the effects of nutrition, epigenetics and mitochondria on aging.[99–101] Indeed, it is likely that mitochondrial epigenetic changes are closely related to aging, and subsequently aging is among the most studied physiological processes in the field of mitochondrial epigenetics. 5-hmC, often considered as a transitional state during demethylation of 5-mC to cytosine, decreases in abundance with aging in mtDNA from the frontal cortex of brain.[102] Aging can also affect the expression of mtDNMT and TET enzymes, which can therefore alter 5-mC and 5-hmC levels in mitochondria.[102] However, despite reports of the alteration of mitochondrial gene expression with aging, the understanding of how aging and aging-related pathologies are associated with mitochondrial epigenetics is at a primitive stage.

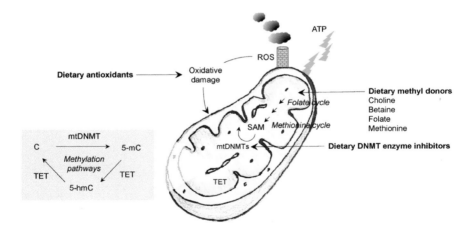

Figure 1. Simplified illustration of potential effects of nutrition on mitochondrial epigenetics. Mitochondria generate ATP and ROS, and oxidative damage to mtDNA caused by ROS may be reduced by dietary antioxidants. Dietary methyl donors can be imported into mitochondria, while mtDNMTs and TETs present in mitochondria may be inhibited by dietary compounds.

There are putative mechanisms by which nutrition may be implicated in this association between aging and mitochondrial epigenetics. Imbalance of nutritional intakes leads to alteration of mitochondrial epigenetic patterns and mitochondrial malfunction that can be related to those associated with the aging process. Such mitochondrial epigenetic changes resulting from the effects of malnutrition or mitochondrial dysfunction can be "drivers" and "passengers" of the aging process.

Toward Understanding of the Complexity of Nutrition- and Age-related Epigenetic Changes

An important feature of the aging process is increasing heterogeneity. This is observed at the whole organism level where individual persons of the same chronological age can have very different physical, cognitive and physiological capabilities.[4] At a cellular level, this heterogeneity is evident as an age-related increase in cell-to-cell variation in gene expression.[41] At an epigenetic level, this heterogeneity is observed as greater inter-individual variation in methylation levels at specific genomic loci with increasing age.[35,38] As we have argued, these time-related changes in methylation at specific loci (and, indeed, in other epigenetic marks) may contribute to the inter-cell heterogeneity within tissues which characterizes the aging process.[40] In summary, over the life course, the cellular makeup of tissues and organs becomes more heterogeneous, resulting in progressively greater inter-cell differences in functional capacity. At the extreme, this is exemplified by the age-dependent increase in the proportion of senescent cells.[103]

Better understanding of the aging process, and of the roles of nutrition and of epigenetic mechanisms in modulating aging, will require:

i. advances in technologies for assessing relevant molecular events at the single-cell level;
ii. conceptual models and associated systems biology approaches to integrate measurements made at the single-cell level to help

explain the consequences for function of tissues, organs and whole organisms.

Advances in high-throughput RNA sequencing (RNA-seq) at the single-cell level have facilitated the large-scale generation of single-cell transcriptomic data but significant computational and analytical challenges remain to be overcome.[104] Similarly, it is becoming possible to interrogate epigenetics marks and molecules at the single-cell level and this is likely to advance through the application of e.g. microfluidics technologies to single-cell isolation and analysis.[105] Understanding age-related changes in epigenetic regulation of gene expression at the single-cell level will be accelerated by new approaches which can map the accessible genome of individual cells through assays for transposase-accessible chromatin using sequencing (ATAC-seq) which have been integrated into a programmable microfluidics platform.[42] Finally, systems are being developed which will allow the integration of data from high-throughput technologies, aiming to provide quantitative biological understanding of aging at cellular and systems levels and which address the heterogeneity which characterizes aging.[106,107] At present, many of these approaches are at a very early stage of development and are being applied in model organisms but we can be optimistic that they will lead to approaches which can be translated to human studies.

References

1. C López-Otín, MA Blasco, L Partridge, M Serrano, G Kroemer. (2013) The hallmarks of aging. *Cell* **153**(6):1194–1217.
2. TB Kirkwood. (2005) Understanding the odd science of aging. *Cell* **120**(4):437–447.
3. AJ Zimm, M Modabber, V Fernandes, K Karimi, PA Adamson. (2013) Objective assessment of perceived age reversal and improvement in attractiveness after aging face surgery. *JAMA Facial Plast Surg* **15**(6):405–410.
4. DW Belsky, A Caspi, R Houts, HJ Cohen, DL Corcoran, A Danese, H Harrington, S Israel, ME Levine, JD Schaefer, K Sugden, B

Williams, AI Yashin, R Poulton, TE Moffitt. (2015) Quantification of biological aging in young adults. *Proc Natl Acad Sci USA* **112**(30):E4104–110.

5. TB Kirkwood, M Feder, CE Finch, C Franceschi, A Globerson, CP Klingenberg, K LaMarco, S Omholt, RG Westendorp. (2005) What accounts for the wide variation in life span of genetically identical organisms reared in a constant environment? *Mech Aging Dev* **126**(3):439–443.
6. J Vijg and Y Suh. (2005) Genetics of longevity and aging. *Annu Rev Med* **56**:193–212.
7. D Gems, L Partridge. (2013) Genetics of longevity in model organisms: debates and paradigm shifts. *Annu Rev Physiol* **75**:621–644.
8. AM Herskind, M McGue, NV Holm, TI Sorensen, B Harvald, and JW Vaupel. (1996) The heritability of human longevity: a population-based study of 2872 Danish twin pairs born 1870–1900. *Hum Genet* **97**(3):319–323.
9. BD Mitchell, WC Hsueh, TM King, TI Pollin, J Sorkin, R Agarwala, AA Schaffer, AR Shuldiner. (2001) Heritability of life span in the Old Order Amish. *Am J Med Genet* **102**(4):346–352.
10. M McGue, JW Vaupel, N Holm, B Harvald. (1993) Longevity is moderately heritable in a sample of Danish twins born 1870–1880. *J Gerontol* **48**(6):B237–244.
11. L Broer, AS Buchman, J Deelen, DS Evans, JD Faul, KL Lunetta, P Sebastiani, JA Smith, AV Smith, T Tanaka, L Yu, AM Arnold, T Aspelund, EJ Benjamin, PL De Jager, G Eirkisdottir, DA Evans, ME Garcia, A Hofman, RC Kaplan, SL Kardia, DP Kiel, BA Oostra, ES Orwoll, N Parimi, BM Psaty, F Rivadeneira, JI Rotter, S Seshadri, A Singleton, H Tiemeier, AG Uitterlinden, W Zhao, S Bandinelli, DA Bennett, L Ferrucci, V Gudnason, TB Harris, D Karasik, LJ Launer, TT Perls, PE Slagboom, GJ Tranah, DR Weir, AB Newman, CM van Duijn, JM Murabito. (2015) GWAS of longevity in CHARGE consortium confirms APOE and FOXO3 candidacy. *J Gerontol A Biol Sci Med Sci* **70**(1):110–118.
12. LA Herndon, PJ Schmeissner, JM Dudaronek, PA Brown, KM Listner, Y Sakano, MC Paupard, DH Hall, M Driscoll. (2002)

Stochastic and genetic factors influence tissue-specific decline in aging C. elegans. *Nature* **419**(6909):808–814.
13. TB Kirkwood, CE Finch. (2002) Aging: the old worm turns more slowly. *Nature* **419**(6909):794–795.
14. JH Hoeijmakers. (2009) DNA damage, aging, and cancer. *N Engl J Med* **361**(15):1475–1485.
15. G Xu, M Herzig, V Rotrekl, CA Walter. (2008) Base excision repair, aging and health span. *Mech Aging Dev* **129**(7–8):366–382.
16. J Tyson, F Caple, A Spiers, B Burtle, AK Daly, EA Williams, JE Hesketh, JC Mathers. (2009) Inter-individual variation in nucleotide excision repair in young adults: effects of age, adiposity, micronutrient supplementation and genotype. *Br J Nutr* **101**(9):1316–1323.
17. F Caple, EA Williams, A Spiers, J Tyson, B Burtle, AK Daly, JC Mathers, JE Hesketh. (2010) Inter-individual variation in DNA damage and base excision repair in young, healthy non-smokers: effects of dietary supplementation and genotype. *Br J Nutr* **103**(11):1585–1593.
18. L Fontana, L Partridge, VD Longo. (2010) Extending healthy life span — from yeast to humans. *Science* **328**(5976):321–326.
19. WR Swindell. (2012) Dietary restriction in rats and mice: a meta-analysis and review of the evidence for genotype-dependent effects on lifespan. *Aging Res Rev* **11**(2):254–270.
20. RJ Colman, RM Anderson, SC Johnson, EK Kastman, KJ Kosmatka, TM Beasley, DB Allison, C Cruzen, HA Simmons, JW Kemnitz, R Weindruch. (2009) Caloric restriction delays disease onset and mortality in rhesus monkeys. *Science* **325**(5937):201–204.
21. JA Mattison, GS Roth, TM Beasley, EM Tilmont, AM Handy, RL Herbert, DL Longo, DB Allison, JE Young, M Bryant, D Barnard, WF Ward, W Qi, DK Ingram, R de Cabo. (2012) Impact of caloric restriction on health and survival in rhesus monkeys from the NIA study. *Nature* **489**(7415):318–321.
22. RJ Colman, TM Beasley, JW Kemnitz, SC Johnson, R Weindruch, RM Anderson. (2014) Caloric restriction reduces age-related and all-cause mortality in rhesus monkeys. *Nat Commun* **5**:3557.
23. C Prospective Studies, G Whitlock, S Lewington, P Sherliker, R Clarke, J Emberson, J Halsey, N Qizilbash, R Collins, R Peto. (2009) Body-mass index and cause-specific mortality in 900 000

adults: collaborative analyses of 57 prospective studies. *Lancet* **373**(9669):1083–1096.
24. Q Sun, MK Townsend, OI Okereke, OH Franco, FB Hu, F Grodstein. (2009) Adiposity and weight change in mid-life in relation to healthy survival after age 70 in women: prospective cohort study. *BMJ* **339**:b3796.
25. C Handschin, BM Spiegelman. (2008) The role of exercise and PGC1alpha in inflammation and chronic disease. *Nature* **454**(7203):463–469.
26. L Fontana, FB Hu. (2014) Optimal body weight for health and longevity: bridging basic, clinical, and population research. *Aging Cell* **13**(3):391–400.
27. World Health Organisation. (2015) *World Health Statistics*, http://www.who.int/gho/publications/world_health_statistics/2015/en/. Accessed 29 July 2015.
28. BJ Willcox, K Yano, R Chen, DC Willcox, BL Rodriguez, KH Masaki, T Donlon, B Tanaka, JD Curb. (2004) How much should we eat? The association between energy intake and mortality in a 36-year follow-up study of Japanese-American men. *J Gerontol A Biol Sci Med Sci* **59**(8):789–795.
29. Y Tomata, T Watanabe, Y Sugawara, WT Chou, M Kakizaki, I Tsuji. (2014) Dietary patterns and incident functional disability in elderly Japanese: the Ohsaki Cohort 2006 study. *J Gerontol A Biol Sci Med Sci* **69**(7):843–851.
30. F Sofi, R Abbate, GF Gensini, A Casini. (2010) Accruing evidence on benefits of adherence to the Mediterranean diet on health: an updated systematic review and meta-analysis. *Am J Clin Nutr* **92**(5):1189–1196.
31. M de Lorgeril, P Salen, JL Martin, I Monjaud, J Delaye, N Mamelle. (1999) Mediterranean diet, traditional risk factors, and the rate of cardiovascular complications after myocardial infarction: final report of the Lyon Diet Heart Study. *Circulation* **99**(6):779–785.
32. R Estruch, E Ros, J Salas-Salvado, MI Covas, D Corella, F Aros, E Gomez-Gracia, V Ruiz-Gutierrez, M Fiol, J Lapetra, RM Lamuela-Raventos, L Serra-Majem, X Pinto, J Basora, MA Munoz, JV Sorli, JA Martinez, MA Martinez-Gonzalez, PS Investigators. (2013)

Primary prevention of cardiovascular disease with a Mediterranean diet. *N Engl J Med* **368**(14):1279–1290.
33. BE Harmon, CJ Boushey, YB Shvetsov, R Ettienne, J Reedy, LR Wilkens, L Le Marchand, BE Henderson, LN Kolonel. (2015) Associations of key diet-quality indexes with mortality in the multiethnic cohort: the Dietary Patterns Methods Project. *Am J Clin Nutr* **101**(3):587–597.
34. JC Kiefte-de Jong, JC Mathers, OH Franco. (2014) Nutrition and healthy aging: the key ingredients. *Proc Nutr Soc* **73**(2):249–259.
35. MF Fraga, E Ballestar, MF Paz, S Ropero, F Setien, ML Ballestar, D Heine-Suner, JC Cigudosa, M Urioste, J Benitez, M Boix-Chornet, A Sanchez-Aguilera, C Ling, E Carlsson, P Poulsen, A Vaag, Z Stephan, TD Spector, YZ Wu, C Plass, M Esteller. (2005) Epigenetic differences arise during the lifetime of monozygotic twins. *Proc Natl Acad Sci USA* **102**(30):10604–10609.
36. RP Talens, K Christensen, H Putter, G Willemsen, L Christiansen, D Kremer, HE Suchiman, PE Slagboom, DI Boomsma, BT Heijmans. (2012) Epigenetic variation during the adult lifespan: cross-sectional and longitudinal data on monozygotic twin pairs. *Aging Cell* **11**(4):694–703.
37. S Lisanti, WA Omar, B Tomaszewski, S De Prins, G Jacobs, G Koppen, JC Mathers, SA Langie. (2013) Comparison of methods for quantification of global DNA methylation in human cells and tissues. *PLoS One* **8**(11):e79044.
38. HE Gautrey, SD van Otterdijk, HJ Cordell, T Newcastle 85+ Study Core, JC Mathers, G Strathdee. (2014) DNA methylation abnormalities at gene promoters are extensive and variable in the elderly and phenocopy cancer cells. *FASEB J* **28**(7):3261–3272.
39. J Collerton, HE Gautrey, SD van Otterdijk, K Davies, C Martin-Ruiz, T von Zglinicki, TB Kirkwood, C Jagger, JC Mathers, G Strathdee. (2014) Acquisition of aberrant DNA methylation is associated with frailty in the very old: findings from the Newcastle 85+ Study. *Biogerontology* **15**(4):317–328.
40. D Ford and JC Mathers. (2009) Nutrition, epigenetics and aging, in *Nutrients and epigenetics*, S-W Choi, S Friso, Editors. CRC Press (Taylor and Francis Group), 175–205.

41. R Bahar, CH Hartmann, KA Rodriguez, AD Denny, RA Busuttil, ME Dolle, RB Calder, GB Chisholm, BH Pollock, CA Klein, J Vijg. (2006) Increased cell-to-cell variation in gene expression in aging mouse heart. *Nature* **441**(7096):1011–1014.
42. JD Buenrostro, B Wu, UM Litzenburger, D Ruff, ML Gonzales, MP Snyder, HY Chang, WJ Greenleaf. (2015) Single-cell chromatin accessibility reveals principles of regulatory variation. *Nature* **523**(7561):486–490.
43. S Bocklandt, W Lin, ME Sehl, FJ Sanchez, JS Sinsheimer, S Horvath, E Vilain. (2011) Epigenetic predictor of age. *PLoS One* **6**(6):e14821.
44. RE Marioni, S Shah, AF McRae, BH Chen, E Colicino, SE Harris, J Gibson, AK Henders, P Redmond, SR Cox, A Pattie, J Corley, L Murphy, NG Martin, GW Montgomery, AP Feinberg, MD Fallin, ML Multhaup, AE Jaffe, R Joehanes, J Schwartz, AC Just, KL Lunetta, JM Murabito, JM Starr, S Horvath, AA Baccarelli, D Levy, PM Visscher, NR Wray, IJ Deary. (2015) DNA methylation age of blood predicts all-cause mortality in later life. *Genome Biol* **16**:25.
45. G Hannum, J Guinney, L Zhao, L Zhang, G Hughes, S Sadda, B Klotzle, M Bibikova, JB Fan, Y Gao, R Deconde, M Chen, I Rajapakse, S Friend, T Ideker, K Zhang. (2013) Genome-wide methylation profiles reveal quantitative views of human aging rates. *Mol Cell* **49**(2):359–367.
46. S Horvath. (2013) DNA methylation age of human tissues and cell types. *Genome Biol* **14**(10):R115.
47. AA Baccarelli, HM Byun. (2015) Platelet mitochondrial DNA methylation: a potential new marker of cardiovascular disease. *Clin Epigenetics* **7**(1):44.
48. S Horvath, P Garagnani, MG Bacalini, C Pirazzini, S Salvioli, D Gentilini, AM Di Blasio, C Giuliani, S Tung, HV Vinters, C Franceschi. (2015) Accelerated epigenetic aging in down syndrome. *Aging Cell* **14**(3):491–495.
49. TM Rickabaugh, RM Baxter, M Sehl, JS Sinsheimer, PM Hultin, LE Hultin, A Quach, O Martinez-Maza, S Horvath, E Vilain, BD Jamieson. (2015) Acceleration of age-associated methylation patterns in HIV-1-infected adults. *PLoS One* **10**(3):e0119201.

50. S Horvath, V Mah, AT Lu, JS Woo, OW Choi, AJ Jasinska, JA Riancho, S Tung, NS Coles, J Braun, HV Vinters, LS Coles. (2015) The cerebellum ages slowly according to the epigenetic clock. *Aging (Albany NY)* **7**(5):294–306.
51. CI Weidner, Q Lin, CM Koch, L Eisele, F Beier, P Ziegler, DO Bauerschlag, KH Jockel, R Erbel, TW Muhleisen, M Zenke, TH Brummendorf, W Wagner. (2014) Aging of blood can be tracked by DNA methylation changes at just three CpG sites. *Genome Biol* **15**(2):R24.
52. LC Greaves, MJ Barron, S Plusa, TB Kirkwood, JC Mathers, RW Taylor, DM Turnbull. (2010) Defects in multiple complexes of the respiratory chain are present in aging human colonic crypts. *Exp Gerontol* **45**(7–8):573–579.
53. CR Reczek, NS Chandel. (2015) ROS-dependent signal transduction. *Curr Opin Cell Biol* **33**:8–13.
54. M Nooteboom, R Johnson, RW Taylor, NA Wright, RN Lightowlers, TB Kirkwood, JC Mathers, DM Turnbull, LC Greaves. (2010) Age-associated mitochondrial DNA mutations lead to small but significant changes in cell proliferation and apoptosis in human colonic crypts. *Aging Cell* **9**(1):96–99.
55. G Bjelakovic, D Nikolova, LL Gluud, RG Simonetti, C Gluud. (2007) Mortality in randomized trials of antioxidant supplements for primary and secondary prevention: systematic review and meta-analysis. *JAMA* **297**(8):842–857.
56. D Bellizzi, P D'Aquila, T Scafone, M Giordano, V Riso, A Riccio, G Passarino. (2013) The control region of mitochondrial DNA shows an unusual CpG and non-CpG methylation pattern. *DNA Res* **20**(6):537–547.
57. V Patil, RL Ward, LB Hesson. (2014) The evidence for functional non-CpG methylation in mammalian cells. *Epigenetics* **9**(6):823–828.
58. H Manev, S Dzitoyeva. (2013) Progress in mitochondrial epigenetics. *Biomol Concepts* **4**(4):381–389.
59. S Das, M Ferlito, OA Kent, K Fox-Talbot, R Wang, D Liu, N Raghavachari, Y Yang, SJ Wheelan, E Murphy, C Steenbergen. (2012) Nuclear miRNA regulates the mitochondrial genome in the heart. *Circ Res* **110**(12):1596–1603.

60. RD Kelly, A Mahmud, M McKenzie, IA Trounce, JC St John. (2012) Mitochondrial DNA copy number is regulated in a tissue specific manner by DNA methylation of the nuclear-encoded DNA polymerase gamma A. *Nucleic Acids Res* **40**(20):10124–10138.
61. D Bellizzi, P D'Aquila, M Giordano, A Montesanto, G Passarino. (2012) Global DNA methylation levels are modulated by mitochondrial DNA variants. *Epigenomics* **4**(1):17–27.
62. FV Duarte, CM Palmeira, AP Rolo. (2014) The Role of microRNAs in Mitochondria: Small Players Acting Wide. *Genes (Basel)* **5**(4):865–886.
63. TL Schwarz. (2013) Mitochondrial trafficking in neurons. *Cold Spring Harb Perspect Biol* **5**(6).
64. S Feng, L Xiong, Z Ji, W Cheng, H Yang. (2012) Correlation between increased ND2 expression and demethylated displacement loop of mtDNA in colorectal cancer. *Mol Med Rep* **6**(1):125–130.
65. HM Byun, N Benachour, D Zalko, MC Frisardi, E Colicino, L Takser, AA Baccarelli. (2015) Epigenetic effects of low perinatal doses of flame retardant BDE-47 on mitochondrial and nuclear genes in rat offspring. *Toxicology* **328**:152–159.
66. BG Janssen, HM Byun, W Gyselaers, W Lefebvre, AA Baccarelli, TS Nawrot. (2015) Placental mitochondrial methylation and exposure to airborne particulate matter in the early life environment: An ENVIRONAGE birth cohort study. *Epigenetics* **10**(6):536–544.
67. HM Byun, T Panni, V Motta, L Hou, F Nordio, P Apostoli, PA Bertazzi, AA Baccarelli. (2013) Effects of airborne pollutants on mitochondrial DNA methylation. *Part Fibre Toxicol* **10**:18.
68. H Chen, S Dzitoyeva, H Manev. (2012) Effect of valproic acid on mitochondrial epigenetics. *Eur J Pharmacol* **690**(1–3):51–59.
69. OH Kramer, P Zhu, HP Ostendorff, M Golebiewski, J Tiefenbach, MA Peters, B Brill, B Groner, I Bach, T Heinzel, M Gottlicher. (2003) The histone deacetylase inhibitor valproic acid selectively induces proteasomal degradation of HDAC2. *EMBO J* **22**(13):3411–3420.
70. S Gu, Y Tian, A Chlenski, HR Salwen, Z Lu, JU Raj, Q Yang. (2012) Valproic acid shows a potent antitumor effect with alteration of DNA methylation in neuroblastoma. *Anticancer Drugs* **23**(10):1054–1066.
71. RK Porter, JM Scott, MD Brand. (1992) Choline transport into rat liver mitochondria. *Biochem Soc Trans* **20**(3):248S.

72. V Michel, M Bakovic. (2012) The ubiquitous choline transporter SLC44A1. *Cent Nerv Syst Agents Med Chem* **12**(2):70–81.
73. SA Craig. (2004) Betaine in human nutrition. *Am J Clin Nutr* **80**(3):539–549.
74. MK Chern, R Pietruszko. (1999) Evidence for mitochondrial localization of betaine aldehyde dehydrogenase in rat liver: purification, characterization, and comparison with human cytoplasmic E3 isozyme. *Biochem Cell Biol* **77**(3):179–187.
75. EA McCarthy, SA Titus, SM Taylor, C Jackson-Cook, RG Moran. (2004) A mutation inactivating the mitochondrial inner membrane folate transporter creates a glycine requirement for survival of chinese hamster cells. *J Biol Chem* **279**(32):33829–33836.
76. YS Shin, C Chan, AJ Vidal, T Brody, EL Stokstad. (1976) Subcellular localization of gamma-glutamyl carboxypeptidase and of folates. *Biochim Biophys Acta* **444**(3):794–801.
77. JW Locasale. (2013) Serine, glycine and one-carbon units: cancer metabolism in full circle. *Nat Rev Cancer* **13**(8):572–583.
78. A Ormazabal, M Casado, M Molero, J Montoya, S Rahman, SB Aylett, I Hargreaves, S Heales, R Artuch. (2015) Can folic acid have a role in mitochondrial disorders? *Drug Discov Today*.
79. AS Tibbetts, DR Appling. (2010) Compartmentalization of Mammalian folate-mediated one-carbon metabolism. *Annu Rev Nutr* **30**:57–81.
80. KD Robertson. (2001) DNA methylation, methyltransferases, and cancer. *Oncogene* **20**(24):3139–3155.
81. SH Choi, K Heo, HM Byun, W An, W Lu, AS Yang. (2011) Identification of preferential target sites for human DNA methyltransferases. *Nucleic Acids Res* **39**(1):104–118.
82. I Bernstein, HM Byun, A Mohrbacher, D Douer, G Gorospe, 3rd, J Hergesheimer, S Groshen, C O'Connell, AS Yang. (2010) A phase I biological study of azacitidine (Vidaza) to determine the optimal dose to inhibit DNA methylation. *Epigenetics* **5**(8):750–757.
83. GS Mack. (2010) To selectivity and beyond. *Nat Biotechnol* **28**(12):1259–1266.
84. MZ Fang, Y Wang, N Ai, Z Hou, Y Sun, H Lu, W Welsh, CS Yang. (2003) Tea polyphenol (-)-epigallocatechin-3-gallate inhibits DNA

methyltransferase and reactivates methylation-silenced genes in cancer cell lines. *Cancer Res* **63**(22):7563–7570.
85. CS Yang, M Fang, JD Lambert, P Yan, TH Huang. (2008) Reversal of hypermethylation and reactivation of genes by dietary polyphenolic compounds. *Nutr Rev* **66 Suppl 1**:S18–20.
86. WJ Lee, JY Shim, BT Zhu. (2005) Mechanisms for the inhibition of DNA methyltransferases by tea catechins and bioflavonoids. *Mol Pharmacol* **68**(4):1018–1030.
87. WJ Lee, BT Zhu. (2006) Inhibition of DNA methylation by caffeic acid and chlorogenic acid, two common catechol-containing coffee polyphenols. *Carcinogenesis* **27**(2):269–277.
88. A Link, F Balaguer, Y Shen, JJ Lozano, HC Leung, CR Boland, A Goel. (2013) Curcumin modulates DNA methylation in colorectal cancer cells. *PLoS One* **8**(2):e57709.
89. Z Liu, S Liu, Z Xie, RE Pavlovicz, J Wu, P Chen, J Aimiuwu, J Pang, D Bhasin, P Neviani, JR Fuchs, C Plass, PK Li, C Li, TH Huang, LC Wu, L Rush, H Wang, D Perrotti, G Marcucci, KK Chan. (2009) Modulation of DNA methylation by a sesquiterpene lactone parthenolide. *J Pharmacol Exp Ther* **329**(2):505–514.
90. KD Sheikh, PP Banerjee, S Jagadeesh, SC Grindrod, L Zhang, M Paige, ML Brown. (2010) Fluorescent epigenetic small molecule induces expression of the tumor suppressor ras-association domain family 1A and inhibits human prostate xenograft. *J Med Chem* **53**(6):2376–2382.
91. Q Xie, Q Bai, LY Zou, QY Zhang, Y Zhou, H Chang, L Yi, JD Zhu, MT Mi. (2014) Genistein inhibits DNA methylation and increases expression of tumor suppressor genes in human breast cancer cells. *Genes Chromosomes Cancer* **53**(5):422–431.
92. LS Shock, PV Thakkar, EJ Peterson, RG Moran, SM Taylor. (2011) DNA methyltransferase 1, cytosine methylation, and cytosine hydroxymethylation in mammalian mitochondria. *Proc Natl Acad Sci USA* **108**(9):3630–3635.
93. M Wong, B Gertz, BA Chestnut, LJ Martin. (2013) Mitochondrial DNMT3A and DNA methylation in skeletal muscle and CNS of transgenic mouse models of ALS. *Front Cell Neurosci* **7**:279.

94. DT Shaughnessy, K McAllister, L Worth, AC Haugen, JN Meyer, FE Domann, B Van Houten, R Mostoslavsky, SJ Bultman, AA Baccarelli, TJ Begley, RW Sobol, MD Hirschey, T Ideker, JH Santos, WC Copeland, RR Tice, DM Balshaw, FL Tyson. (2014) Mitochondria, energetics, epigenetics, and cellular responses to stress. *Environ Health Perspect* **122**(12):1271–1278.

95. AW Ashor, M Siervo, J Lara, C Oggioni, S Afshar, JC Mathers. (2015) Effect of vitamin C and vitamin E supplementation on endothelial function: a systematic review and meta-analysis of randomised controlled trials. *Br J Nutr* **113**(8):1182–1194.

96. OJ Switzeny, E Mullner, KH Wagner, H Brath, E Aumuller, AG Haslberger. (2012) Vitamin and antioxidant rich diet increases MLH1 promoter DNA methylation in DMT2 subjects. *Clin Epigenetics* **4**(1):19.

97. JA Colacino, AE Arthur, DC Dolinoy, MA Sartor, SA Duffy, DB Chepeha, CR Bradford, HM Walline, JB McHugh, N D'Silva, TE Carey, GT Wolf, JM Taylor, KE Peterson, LS Rozek. (2012) Pretreatment dietary intake is associated with tumor suppressor DNA methylation in head and neck squamous cell carcinomas. *Epigenetics* **7**(8):883–891.

98. M Garcia-Lacarte, FI Milagro, MA Zulet, JA Martinez, ML Mansego. (2015) LINE-1 methylation levels, a biomarker of weight loss in obese subjects, are influenced by dietary antioxidant capacity. *Redox Rep*.

99. BA Payne, PF Chinnery. (2015) Mitochondrial dysfunction in aging: Much progress but many unresolved questions. *Biochim Biophys Acta*.

100. JC Mathers. (2015) Impact of nutrition on the aging process. *Br J Nutr* **113 Suppl**:S18–22.

101. JA McKay, JC Mathers. (2011) Diet induced epigenetic changes and their implications for health. *Acta Physiol (Oxf)* **202**(2):103–118.

102. S Dzitoyeva, H Chen, H Manev. (2012) Effect of aging on 5-hydroxymethylcytosine in brain mitochondria. *Neurobiol Aging* **33**(12):2881–2891.

103. DJ Baker, T Wijshake, T Tchkonia, NK LeBrasseur, BG Childs, B van de Sluis, JL Kirkland, JM van Deursen. (2011) Clearance of

p16Ink4a-positive senescent cells delays aging-associated disorders. *Nature* **479**(7372):232–236.
104. O Stegle, SA Teichmann, JC Marioni. (2015) Computational and analytical challenges in single-cell transcriptomics. *Nat Rev Genet* **16**(3):133–145.
105. P Bheda, R Schneider. (2014) Epigenetics reloaded: the single-cell revolution. *Trends Cell Biol* **24**(11):712–723.
106. EC Spivey, IJ Finkelstein. (2014) From cradle to grave: high-throughput studies of aging in model organisms. *Mol Biosyst* **10**(7):1658–1667.
107. SM Jazwinski, AI Yashin. (2015) Aging and health — a systems biology perspective. Introduction. *Interdiscip Top Gerontol* **40**:VII–XII.

Nutrition and Epigenetics: Evidence for Multi- and Transgenerational Effects

CHAPTER 7

Cheryl S. Rosenfeld

Genetics Area Program,
Thompson Center for Autism and Neurobehavioral Disorders,
Bond Life Sciences Center, Department of Biomedical Sciences,
University of Missouri–Columbia,
E102 Veterinary Medical Building, 1600 Rollins,
Columbia, MO 65211, USA

Introduction

To gauge whether various nutritional changes indeed lead to multi- or transgenerational effects, it is important to define each of these terms up front, along with the related concept of developmental origins of health and disease. Maternal and paternal experiences prior to conception through to lactation, in the case of the mother, can dramatically sculpt future offspring health. This idea was originally conceptualized by the late Sir David Barker, leading to its coinage of the "Barker Hypothesis". It then mutated to be called "fetal origin of adult disease (FOAD)", and the final term that has been settled upon is "developmental origin of health and disease (DOHaD)".[1,2] In essence, a DOHaD effect occurs in response to a change in either the maternal or paternal environment that alters conceptus/perinatal programming and subsequent progeny health for better or worse. A paternal (P_0) DOHaD-effect could be attributed to changes in the epigenome of his germ cells and/or his reproductive tract secretions that directly or indirectly impact his F_1

sons and daughters.[3] The time periods and possible mechanisms by which a change in maternal or paternal nutritional status can result in offspring DOHaD alterations are summarized in Table 1.

Table 1. Time periods and mechanisms by which maternal and paternal may lead to offspring DOHaD effects.

Time period	Mom	Dad
Periconceptional (Prior to Fertilization)	Disruptions in oocyte maturation	Epigenetic/phenotypic disruptions in spermatozoa development
Periconceptional (Prior to fertilization)		Small RNA changes in spermatozoa that may occur during spermatogenesis or after transition in the epididymis
Periconceptional (Prior to fertilization) or Early Embryonic Period		Changes in seminal fluid composition that directly impacts zygotic development or interacts with the female reproductive system to affect embryo development
In Utero Period	Alterations in the intra-uterine environment that may affect placenta function or fetal development	
Postnatal Period	Alteration in milk composition	
Postnatal Period	Compromised maternal care	Compromised paternal care, especially in species who are monogamous and biparental

While some DOHaD phenotypes may be propagated to future generations (transgenerational propagation), other changes may arrest at the F_1 generation.

A multigenerational effect can be considered as one where the parental (P_0) lineages, F_1, F_2, and future descendants are subjected to identical environment conditions as the original founders, such as a high fat diet or other nutritional alteration. However, the phenotype, for instance obesity, becomes progressively more pronounced across generations. In contrast, a transgenerational effect is one where the original insult is assumingly removed after the F_1 generation.

To verify transgenerational transmission, it is essential to carry out the study through the F_3 or, even better yet, the F_4 generation.[4,5] Besides impacting the P_0 and F_1 generations, a periconceptional-*in utero* environmental change will also influence the germ cells of the F_1 generation that contribute to the F_2 generation. Thus, a single

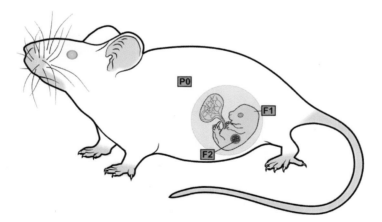

Figure 1. Illustration of how a paternal or maternal (P_0) nutritional change during the periconception and/or gestational period might impact the developing F_1 offspring, including the primordial germ cells, which will give rise to the F_2 generation. Thus, three generations (P_0 through F_2) can be targeted with a single dietary change. Consequently, any study claiming multi- or transgenerational effects must be followed through to the F_3 and possibly F_4 generations. Figure modified from.[84]

"perturbation" to the mother or father can simultaneously affect three generations (P_0 through F_2) (Figure 1).

Moreover, later parental behaviors may be vulnerable to *in utero* environmental changes.[6] Poor parental care by F_1 mothers and fathers may detrimentally shape the epigenome and programming of their F_2 pups. With these caveats in mind, we will consider the evidence to date that various nutritional changes can lead to multi- and transgenerational effects.

Methyl Enriched and Deficient Diets

Pregnant women and those seeking to become pregnant are increasingly being advised to consume prenatal vitamins that contain high amounts of folic acid (vitamin B_9) and methyl supplements, including methionine, pyridoxine (vitamin B_6), betaine (trimethylglycine), choline, and vitamin B_{12} (Figure 2), to prevent neural-tube and other central nervous system (CNS) defects in their offspring.

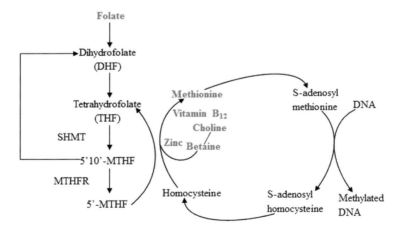

Figure 2. Methyl supplements and methyl activation cycle. Methyl groups can originate from the folate cycle or from betaine. Methyl groups can then be added to cytosine residues of DNA, especially in CpG rich regions, or used to modify histone proteins. Vitamin B_{12} and zinc serve as cofactors for the reactions. Compounds in red font are commonly employed in methyl supplemented diets for rodent model studies. Figure modified from.[12]

Current recommendations are consumption of ~0.8 mg of folic acid one month prior to conception and throughout the first few months of pregnancy.[7–10] While such supplements have greatly reduced the incidence of these congenital disorders, there is now mounting concern that excessive intake of such nutriceuticals may also result in deleterious health consequences in exposed offspring and possibly their descendants. The reason is that many of these compounds can act as methyl donors in the methyl cycle or serve as cofactors in these pathways (Figure 2).

The net effects are to add methyl groups to cytosine residues, especially in CpG rich areas within the promoter or coding region, and histone proteins. Both of these are considered epigenetic changes in that they can alter the expression of a gene but not the DNA itself.[11] Importantly, such epigenetic changes are indiscriminate. Thus, nutritional tampering with the epigenome can simultaneously impact "beneficial" and "harmful" genes. The pendulum to health or disease may depend on the cumulative changes induced by methyl supplemented (MS) diets. The DOHaD effects of such diets has been previously reviewed.[12] Here, we will focus on the evidence that MS and methyl deficient diets can exert multi- or transgenerational effects.

Methyl Supplemented (MS) Diets

The two studies examining the interaction of MS diets and transgenerational changes have employed the viable yellow (A^{vy}/a) mouse model. These mice demonstrate a range of coat color patterns varying from all "brown" (pseudoagouti) to all yellow with intermediate mottled pelage patterns also occurring. In rodent species, the *agouti* (*A*) gene encodes for the agouti signaling protein (ASIP), whose expression is confined to the mid-phase of the hair cycle and results in the deposition of yellow/red (pheomelanin) pigments.[13] Therefore, the hair follicle has a combination of eumelanin (brown/black) pigments and pheomelanin. In A^{vy}/a mice, a retrotransposon (intracisternal A particle-IAP) has evolutionary inserted in pseudoexon 1a, and this viral insert possesses its own cryptic promoter site that usurps transcriptional control of *A*, potentially resulting in its systemic and constitutive expression.[14,15] Yet, the viral particle-induced expression of this gene is highly variable, which leads to these mice being epigenetic mosaics. The coat color range reflects the varying degree of methylation and activation state of the A^{vy}/a allele during fetal development. The fur of those possessing a hypomethylated (active) viral promoter region is generally yellow; whereas, hypermethylation (inactivation of the viral promoter site) results in a pseudoagouti coat color. Multicolored mice are due to mosaicism representing the number of clonal patches of tissue where the A^{vy} allele is either active (demethylated) or inactive (methylated). Yellow coat color A^{vy}/a mice, especially males, are predisposed to developing "yellow obese syndrome" typified by hyperglycemia, obesity, and increased appetite.[16,17] The increased appetite and obesity is presumably due to ectopic expression of *agouti* in the hypothalamus,[18] where it antagonizes the melanocortin 4 receptor (MC4R),[13] and pancreas.[19]

When the A^{vy}/a allele is passed through the maternal germline, there appears to be incomplete erasure of the methylation marks on the IAP site, such that yellow coat color mothers birth a preponderance of offspring with the same coat color as the mother. This effect persists to her grand-offspring (F_2 generation).[15] Similar transgenerational effects are not observed when this allele is inherited through the paternal germline. Thus, the reprogramming of the IAP methylation state in F_1 offspring derived from A^{vy}/a males is random, along with the resulting coat color patterns.

A maternal MS diet to *a/a* (black) C57Bl6J mice, gestating A^{vy}/a conceptuses, enhances the average methylation state of the IAP and skews the proportion of offspring towards the browner end of the color coat spectrum. Thus, developmental exposure to a MS can induce positive health benefits by reducing the percentage of yellow coat color offspring who will go on to develop later metabolic disorders.[20–23] Based on the results, a follow-up study determined that transmission of the A^{vy}/a allele through the female germline led to amplification of obesity in successive generations.[24] However, maternal MS reversed this transgenerational effect on body weight but without altering the epigenetic state of the A^{vy} gene. The findings suggest that other genes might be influenced by this maternal diet, and such effects could potentially persist for several generations. Yet, there does not appear to be any cumulative multigenerational benefits of feeding a MS diet across generations to A^{vy}/a females.[25] In other words, F_2 and F_3 generations derived from F_1 and F_2 A^{vy}/a dams, respectively, fed a MS diet do not demonstrate greater percentage of darker coat color offspring relative to the F_1 generation whose mothers were provisioned with this same supplemented diet. Therefore, the scant data to date suggests there might be select transgenerational effects of a MS diet that may confer some positive health benefits. However, it remains to be determined how ancestral exposure to such supplements impacts other rodent models and humans.

Methyl Deficient Diets

Developmental exposure to methyl deficient diets is associated with a variety of health issues. Rat offspring whose mothers were provided a folate and vitamin B_{12} deficient diet during gestation and lactation demonstrate decreased synapsin expression in the cerebellum that is attributed to impairment of estrogen receptor 1 (ESR1) signaling.[26] Deficiency of folic acid in late gestation suppresses progenitor cells proliferation but increases apoptosis in the fetal mouse brain.[27] Neurobehavioral programming is also vulnerable to folate deficiency after weaning, as evidenced by the fact that learning and memory deficits are reported in rats placed on a folate deficient diet at weaning but who are supplied normal folate concentrations during the perinatal period.[28] Maternal folate and vitamin B_{12} deficient diet cause gene expression changes for other neurotrophins and signaling molecules in the brain and DNA methylation patterns,[29] which may be improved with omega-3 fatty acid supplementation.[30] Concurrent omega-3 fatty acid supplementation may also improve the effects of maternal vitamin B_{12} or folate deficiency on placental gene expression,[31] milk composition in the dams subjected to this diet regimen,[32] and female offspring reproductive cycles.[33]

Other fetal organs are vulnerable to maternal folate and/or vitamin B_{12} deficiency. A low folate and selenium diet fed to mouse mothers during gestation and lactation alters hepatic expression in their offspring for genes regulating several metabolic pathways and methyl group metabolism.[34] Glucose homeostasis is disrupted in male but not female offspring born to rat dams fed a methyl deficient diet during the periconception period.[35] Administration of maternal folic acid restricted or methyl-donor deficient diets to rats disturbs offspring lipid metabolism[36,37] and insulin regulation.[38] It may be hypothesized that the phenotypic changes are due to altered DNA methylation patterns. However, this maternal diet regimen does not change global patterns of DNA methylation in the fetal liver, even though prevalent changes in choline and amino acid were evident.[39] Both maternal and paternal folate deficiency appears to affect rat offspring global DNA methylation patterns and protein expression of FRA, IGF2, and IGF1R in the liver.[40] Paternal folate deficiency may also impact IGF2 protein expression and DNA methylation patterns in the brain, even when dams are provided sufficient folate concentrations.[41] Male mice placed at weaning on a folate deficient diet exhibit decreased cauda epididymal sperm numbers, increase DNA fragmentation index, and increased expanded simple tandem repeats (ESTR).[42] These and potential epigenetic changes may account for later offspring DOHaD effects and even possible transgenerational propagation.

Maternal vitamin B_{12} or folate deficient diets may underpin proteomic changes (e.g. up-regulation of calreticulin, cofilin 1, and nucleotide diphosphate kinase B) in the kidneys of rat offspring, although the effects are heightened in the former group.[43] Piglets born to sows provided from early to mid-gestation a folate deficient diet demonstrate changes in body weight, myogenesis in the logissimus dorsi muscle, intra-myofiber triglyceride content, and transcriptomic profiles in skeletal muscle.[44] Ocular malformations have been observed in offspring mice born to mothers maintained on a folic acid deficient diet.[45] Peri/postnatal exposure of rats to a folate and/or vitamin B_{12} deficient diet increases visceral adiposity and disrupts lipid metabolism, which may collectively predispose them to later metabolic diseases.[46] While definitive results have yet to be ascribed to vitamin B_{12} deficient diets in humans, it is apparent that maternal/offspring vitamin B_{12} deficiency is a significant health concern in certain countries, including Mexico and Guatemala.[47,48] Even though several studies have considered whether deficiencies in these micronutrients can result in harmful DOHaD effects, to the author's knowledge only a single study to date has examined whether such diets result in multi- or transgenerational effects. This study placed mice P_0 mothers and their F_1 offspring on a methyl deficient diet and reported no effect on mutation rate for an ESTR locus. Thus, the authors concluded that there was no cause for concern for multigenerational propagation.[49] Given the ample evidence that maternal and paternal methyl deficient diets can influence the health of their sons and daughters, further research is clearly needed to determine whether such diets can induce multi- or transgenerational effects.

Hypercaloric/High Fat Diets

To date, select studies claim to find transgenerational effects in response to a high caloric diet. One study with *Drosophilia* suggests that maternal caloric excess (high sugar diet) results in an obese-like phenotype and altered metabolic genes in two generations of larval offspring subjected to the same high caloric diet.[50] As the adult offspring are placed on the same diet as the founder female, the observed changes are multi- rather than transgenerational. Another study with *Drosophilia* larvae maintained on a high protein/low sugar larval diet report that these offspring metamorphosize at a faster rate, show improvements in reproductive function, and metabolic changes relative to those whose parents were provided a low protein relative to sugar diet.[51]

A maternal hypercaloric diet to P_0 rat dams increases body weight gain, retroperitoneal fat weight, induces hypertrophy in hypodermic adipocytes, but decreases fibrillary acidic protein (GFAP) in

F_1 offspring, which the authors suggest are evidence for transgenerational transmission.[52] Yet, the pathologies are observed in F_1 offspring, who would have been exposed to this nutritional insult while *in utero*. Thus, the phenotypic changes may be ascribed to DOHaD alterations but are not necessarily transgenerational in origin.

The potential for a maternal or paternal high fat diet (HFD) to modulate multi- and transgenerational changes has been explored in a handful of reports. Provisioning P_0 female mice with a HFD stimulates obesogenic and diabetogenic pathologies in their F_1 offspring.[52] However, *in utero* exposure to a control diet in the next three generations of descendants is able to ablate these ancestral changes. The findings imply that maternal induced disruptions may be reversible, and thus a maternal HFD might not result in permanent transgenerational disease outcomes. Conversely, another study concluded that feeding of a HFD over three consecutive generations to mice induces earlier onset and increases severity of obesity from the P_0 to F_2 generation (multigenerational effects), especially in males whose mothers and grandmothers became obese.[53] This group of males also show greater amounts of hepatic steatosis lipogenesis, endoplasmic reticulum stress, histone protein modifications, and elevations in insulin and leptin concentrations. Feeding a Western HFD (35% energy as fat) to male and female mice over four generations results in progressive adipocyte hypertrophy and hyperplasia, hyperinsulinemia, and alterations in *Csf3* and *nocturnin* in the stroma vascular of adipose tissue.[54] P_0 rat dams maintained on an energy-rich diet leads to progressive elevation in *Dnmt3a2* expression and decrease DNA methylation of this promoter site in F_3 compared to F_1 descendants.[55]

There is select evidence that both maternal and paternal HFD can lead to transgenerational effects. A maternal HFD in mice leads to an increase body size and suppression of insulin sensitivity, and these phenotypic changes are transmitted in some cases via the maternal and in other cases through paternal germ cell lines to the next two generations.[56] Further examination of this phenomenon reveals that in F_3 females the increased body size is only conferred through the paternal germline.[57] Additionally, the livers of these females possess changes in paternally expressed transcripts. Together, the findings implicate disturbances in the normal programming of imprinted genes that may contribute to transgenerational transmission, but such alterations appear to be sex-specific. Male and female F_2 descendants derived from male mice provided a HFD that stimulates obesity but in the absence of diabetes are sub-fertile, and this reproductive phenotype is transmitted via the F_1 paternal line to both F_2 sexes and through the F_1 maternal line to F_2 males.[58] A subsequent study by this group revealed that this paternal feeding regimen initiates transgenerational transmission of obesity and insulin resistance through the F_2 generation, which is transmitted via both F_1 parental lineages.[59] Coding and non-coding (micro-RNAs — miRs) are altered in the testes of the P_0 HFD males, who also exhibit germ cell DNA methylation changes. However, when these males are subjected to diet restriction and/or exercise, improvements are noted in their metabolic status and sperm structure and function, suggesting that intervention strategies may be helpful in improving the health of obese males, their offspring, and possibly generations to come.[60]

The collective studies with rodent and *Drosophilia* models provide some evidence that a maternal or paternal HFD or caloric rich diet can result in epigenetic and metabolic changes across the generations. Yet, some of the effects may be alleviated by placing later generations on a control diet, which argues against permanent transgenerational changes.

Malnutrition

The best evidence that parental nutritional status can result in transgenerational effects comes from the Dutch hunger famine that spanned from 1944 to 1945 when Nazi Germany blocked shipment of food into this region. Based on well-documented birth records and interviews, several studies report on transgenerational effects in F_2 grand-offspring. One such report showed that F_1 women exposed *in utero* to famine conditions gave birth to F_2 sons and daughters with increased neonatal adiposity and greater morbidity later in life.[61] In contrast, another study found that F_2 adult offspring whose F_1 fathers were exposed to these under-nourishing conditions during *in utero* life exhibited increased body weight and BMI, but similar effects were not observed in descendants whose F_1 mothers were nutritionally deprived during this early period.[62]

The birth records kept by the parish of Överkalix in northern Sweden from the late 1800s to the 1900s has allowed for assessments on how grandparental nutritional state may impact their grand-offspring. The collective studies show that while grand-maternal state is associated with increase disease conditions in her granddaughters, food availability of paternal grandfathers is linked with obesity and cardiovascular disease in his male descendants.[63–65]

Transgenerational changes in the hypothalamic-pituitary-axis (HPA) and cardiometabolic system may originate in pregnant P_0 guinea pigs fed 70% of their normal ration either early or late in gestation.[66] Feeding this diet late in gestation suppresses birth weight and growth, alters HPA responsiveness, and raises cortisol in the F_1 and to a greater extent in F_2 descendants. On the other hand, cardiac hypertrophy is evident in F_1 and F_2 males derived from P_0 dams subjected to caloric-restriction during early gestation. Hyperglycemia and hyperinsulinemia occurs in F_3 great grandsons of rat dams undernourished either early or late in gestation.[67]

In *Caenorhabditis elegans*, starvation triggers developmental arrest and initiation of small RNA changes that are transgenerationally propagated to F_3 descendants of starved individuals, who also demonstrate an increased lifespan.[68] Small RNAs (to be discussed later) might be one method for one generation to confer memory of past environmental state and allow for subsequent generations to adapt to these historic conditions. The problem arises though when circumstances fluctuate across generations such that future descendants have been prepared by their ancestors to anticipate starvation and instead are exposed to a nutritional surfeit, as occurred after World War II.

It has been proposed that many ethnic groups carry a so-called "thrifty" genotype.[69,70] This concept stipulates that populations repeatedly exposed to food deprivation are prepared to anticipate and adapt to starvation periods, even during *in utero* life. It also suggests that when such individuals are confronted with improved dietary conditions, they are unable to rein in the thrifty genes and so metabolic balance is tipped towards caloric excess and poor control of appetite. It should be emphasized, however, that no thrifty genes have been identified in humans, and proof for the concept is still lacking. To model further how nutritional deprivation might result in transgenerational effects, two primary nutritional deficiencies (protein and vitamin D) have been employed in the laboratory setting.

Protein Restricted Diets

The F_1 through F_3 rat descendants of P_0 rat dams subjected to a protein restricted diet exhibit hepatic gene expression changes.[71] However, the only transcript altered in all generations is one mediating the "Adherens Junctions" pathway. Thus, it raises questions as to whether the effects are indeed transgenerational or variable across generations.

Another consideration is the period when P_0 dams are exposed to protein restricted conditions may lead to differential effects in later generations. At birth, F_1 female rat pups from gestationally protein restricted dams weigh less than controls.[72] F_1 females whose mothers were either fed a protein restricted diet only during gestation or during gestation and lactation exhibit decrease body weight and food intake but increase insulin sensitivity. F_2 males of founder dams protein restricted during lactation are insulin resistant; whereas protein restriction of P_0 females during gestation results in insulin resistance in F_2 female descendants. A similar maternal diet deprivation approach during gestation and lactation reveals glucose and insulin metabolism disturbances in male and female F_2 mice offspring.[73] However, sex differences in windows of vulnerability are observed with F_2 males more affected by ancestral gestational exposure. In contrast, F_2 females are more vulnerable when their grandmother is protein restricted during lactation. A similar protein restriction study in mice, however, concluded that glucose homeostasis is maintained in F_1 through F_3 decedents due to endocrine pancreas adaptation and enhanced sensitivity to circulating concentrations of insulin during the perinatal period.[74]

Gestational protein restriction of P_0 rat dams is associated with hypertension, inhibition of normal parasympathetic-induced vasodilation, and changes in NO production in their F_2 and F_3 male descendants.[75] It cannot be determined though whether the potential transgenerational effects are maternally or paternally inherited as exposed F_1 and F_2 males and females were paired together.

Vitamin D Deficient Diets

One study investigated how maternal vitamin D restriction affects metabolic parameters in F_1 and F_2 rats.[76] While the authors conclude that this maternal diet led to transgenerational changes, phenotypic effects were only evident in F_1 offspring, and thus the data instead support disruptions in DOHaD programming. The phenotypic changes observed in this generation include increase body mass, insulin secretion, area under the curve (AUC) in an oral glucose tolerance test, severe hepatic steatosis, and heightened expression of fatty acid synthase (FAS) protein. This same maternal diet treatment, however, results in rapid onset and more severe experimental autoimmune encephalomyelitis (EAE, an experimental disease condition designed to model multiple sclerosis in humans) in the F_2 generation.[77] It is difficult to determine though whether such F_2 effects are indeed transgenerational in origin or due to other confounding factors detailed above.

Collectively, the current rodent model data implicate that maternal undernourishment, especially for protein, might lead to transgenerational effects, but final outcomes depend upon sex of the descendants, when the nutritional challenge to founder dams occurs, i.e. during gestation or lactation, and rodent model tested. As with other maternal/paternal nutritional changes, it is important that additional studies are carried through to at least the F_3 generation. Cross-fostering approaches may be useful in controlling for potential parental disruptions in the exposed F_1 generation.

Conclusions and Future Directions

The current studies are insufficient to draw definitive conclusions as to whether maternal and paternal diets result in multi- and transgenerational effects. Many of the reports that claim such changes are misleading as the experiments were only followed through to the F_1 and/or F_2 generations. Epigenetic or phenotypic changes observed in F_1 offspring may be attributed to DOHaD programming during the periconception (in the case of paternal or maternal diets), gestational, and/or lactational periods. A single maternal or paternal diet perturbation can concurrently affect three generations, as the F_1 germ cells that will give rise to the F_2 generation are concurrently exposed (Figure 1). Disrupted parenting responses by affected F_1 mothers and possibly fathers might also modulate changes in the F_2 generation. Only a few of the diet studies discussed carried the experiments through to the F_3 or F_4 generation. The reports that have done so yield conflicting findings. In some instances, the changes observed in the later generations differ from those present in earlier generations. Others suggest that placement on a control diet or correction of nutritional deficiencies in later generations reverses disturbances detected in early generations. A true diet-induced transgenerational effect should persist at least through the F_3 generation and possibly beyond. The aforementioned studies revealing that continual placement across generations on a supplemented or restricted diet leads to speedier and exacerbated phenotypic changes support the notion that nutritional status can underpin multigenerational changes. This is of particular concern as high fat diets that are obesogenic and diabetogenic are becoming the mainstay of Western countries. It is currently impossible to determine how such diets might impact our descendants. However, the animal model studies suggest that there is cause for concern, especially for multigenerational changes.

The World Health Organization (WHO) estimates that close to 350 million people worldwide already have type 2 diabetes and that this number is increasing annually, particularly in the developing countries as populations gain greater access to a so-called Western-style diet and become more sedentary.[78] There is no foreseeable decrease in our consumption of such diets. Therefore, it is important that human epidemiological and animal model studies are initiated to examine how our current exposure to such diets might impact our children and future descendants, who may be exposed to even greater caloric-rich diets.

The other critical issue is establishing the potential epigenetic inheritance mechanisms resulting in multi- and transgenerational effects. In the past, the primary mechanism postulated for the perpetual responses was altered gene expression due to changes in chromatin state that may occur due to specific DNA methylation and histone protein modifications. However, such possible mechanisms raise the issue of how such information can be continually transmitted across the male or female germ line and reconstituted to appear in descendants. This possibility is especially troubling given the degree of de-methylation that occurs during early embryonic development.[79,80] Additional evidence against DNA methylation contributing to this "epigenetic memory" comes from organisms who do not methylate DNA to any significant degree, for example the nematode- *C. elegans*. Even so, in response to environmental changes, this organism is able to transmit heritable epigenetic changes, such as through histone methylation footprint and piwi-interacting (pi) and other small RNAs, in response to environmental challenges.[68,81–83] It is also doubtful that histone marks can remain constant across generations and be the purveyors of transgenerational communication. The most plausible candidates for intergenerational propagation are RNA molecules, e.g. miRs, nc-RNA, or piRNA. These biomolecules may selectively bind DNA or coding RNAs and suppress transcription or translation of select genes or class of genes, as has been reported in transgenerational studies with *C. elegans*.[68,81–83]

In summary, there is mounting interest in whether maternal and paternal dietary changes may induce multi- and transgenerational effects. The current results provide some hints that changes in maternal and paternal nutritional state can result in multigenerational effects, but the evidence for transgenerational effects is lacking. The primary reason for this deficiency is that in many cases the animal model studies failed to follow sufficient generations of descendants not directly exposed to the nutritional perturbation. Yet, there is regrettably some evidence for damaging transgenerational effects in individuals whose ancestors were exposed to the "Dutch Hunger Winter" from 1944–1945. Future animal model studies need to address these critical questions: (1) Which nutritional insults result in deleterious multi- and transgenerational changes? (2) Do potential interactions exist between *when* the founder parents are exposed to the nutritional insult and sex of their descendants in terms of vulnerability to such changes, and (3) Which organs systems are most susceptible to multi- and transgenerational changes? Once we have this information in hand, we can then begin to delineate the underlying, presumably epigenetic, mechanisms.

References

1. M Hanson. (2015) The birth and future health of DOHaD. *J Dev Orig Health Dis*:1–4.
2. DJ Barker. (2007) The origins of the developmental origins theory. *J Intern Med* **261**:412–417.
3. OJ Rando, RA Simmons. (2015) I'm eating for two: parental dietary effects on offspring metabolism. *Cell* **161**:93–105.
4. NC Whitelaw, E Whitelaw. (2006) How lifetimes shape epigenotype within and across generations. *Hum Mol Genet* **15** Spec No 2:R 131–137.
5. NA Youngson, E Whitelaw. (2008) Transgenerational epigenetic effects. *Annu Rev Genomics Hum Genet* **9**:233–257.
6. SA Johnson, AB Javurek, MS Painter, MP Peritore, MR Ellersieck, RM Roberts, CS Rosenfeld. (2015) Disruption of parenting behaviors in

California Mice, a monogamous rodent species, by endocrine disrupting chemicals. *PLoS One* **10**:e0126284.
7. http://www.uspreventiveservicestaskforce.org/uspstf/uspsnrfol.htm.
8. T Wolff, CT Witkop, T Miller, SB Syed. (2009) Folic acid supplementation for the prevention of neural tube defects: an update of the evidence for the U.S. Preventive Services Task Force. Rockville MD.
9. T Wolff, CT Witkop, T Miller, SB Syed. (2009) Folic acid supplementation for the prevention of neural tube defects: an update of the evidence for the U.S. Preventive Services Task Force. *Annals of Internal Medicine* **150**:632–639.
10. J Steenblik, E Schroeder, B Hatch, S Groke, C Broadwater-Hollifield, M Mallin, M Ahern, T Madsen. (2011) Folic acid use in pregnant patients presenting to the emergency department. *Int J Emerg Med* **4**:38.
11. CS Rosenfeld. (2010) Animal models to study environmental epigenetics. *Biol Reprod* **82**:473–488.
12. RJ O'Neill, PB Vrana, CS Rosenfeld. (2014) Maternal methyl supplemented diets and effects on offspring health. *Front Genet* **5**:289.
13. RD Cone, D Lu, S Koppula, DI Vage, H Klungland, B Boston, W Chen. DN Orth, C Pouton, RA Kesterson. (1996) The melanocortin receptors: agonists, antagonists, and the hormonal control of pigmentation. *Recent Prog Horm Res* **51**:287–318.
14. EJ Michaud, SJ Bultman, ML Klebig, MJ van Vugt, LJ Stubbs, LB Russell, RP Woychik. (1994) A molecular model for the genetic and phenotypic characteristics of the mouse lethal yellow (Ay) mutation. *Proc Natl Acad Sci USA* **91**:2562–2566.
15. HD Morgan, HG Sutherland, DI Martin, E Whitelaw. (1999) Epigenetic inheritance at the agouti locus in the mouse. *Nat Genet* **23**:314–318.
16. GL Wolff. (1997) Obesity as a pleiotropic effect of gene action. *Journal of Nutrition* **127**:1897S–1901S.
17. GL Wolff, DW Roberts, DB Galbraith. (1986) Prenatal determination of obesity, tumor susceptibility, and coat color pattern in viable yellow (Avy/a) mice. The yellow mouse syndrome. *J Hered* **77**:151–158.
18. SJ Bultman, EJ Michaud, RP Woychik. (1992) Molecular characterization of the mouse agouti locus. *Cell* **71**:1195–1204.

19. M Mansour, D White, C Wernette, J Dennis, YX Tao, R Collins, L Parker, E Morrison. (2010) Pancreatic neuronal melanocortin-4 receptor modulates serum insulin levels independent of leptin receptor. *Endocrine* **37**:220–230.
20. CA Cooney, AA Dave, GL Wolff. (2002) Maternal methyl supplements in mice affect epigenetic variation and DNA methylation of offspring. *J Nutr* **132**:2393S–2400S.
21. RA Waterland, RL Jirtle. (2003) Transposable elements: targets for early nutritional effects on epigenetic gene regulation. *Mol Cell Biol* **23**:5293–5300.
22. GL Wolff, RL Kodell, SR Moore, CA Cooney. (1998) Maternal epigenetics and methyl supplements affect agouti gene expression in Avy/a mice. *FASEB J* **12**:949–957.
23. JE Cropley, CM Suter, KB Beckman, DI Martin. (2010) CpG methylation of a silent controlling element in the murine Avy allele is incomplete and unresponsive to methyl donor supplementation. *PLoS One* **5**:e9055.
24. RA Waterland, M Travisano, KG Tahiliani, MT Rached, S Mirza. (2008) Methyl donor supplementation prevents transgenerational amplification of obesity. *Int J Obes (Lond)* **32**:1373–1379.
25. RA Waterland, M Travisano, KG Tahiliani. (2007) Diet-induced hypermethylation at agouti viable yellow is not inherited transgenerationally through the female. *FASEB J.*
26. G Pourie, N Martin, C Bossenmeyer-Pourie, N Akchiche, RM Gueant-Rodriguez, A Geoffroy, E Jeannesson, H Chehadeh Sel, K Mimoun, P Brachet, V Koziel, JM Alberto *et al.* (2015) Folate- and vitamin B12-deficient diet during gestation and lactation alters cerebellar synapsin expression via impaired influence of estrogen nuclear receptor alpha. *FASEB J* **29**:3713–3725.
27. CN Craciunescu, EC Brown, MH Mar, CD Albright, MR Nadeau, SH Zeisel. (2004) Folic acid deficiency during late gestation decreases progenitor cell proliferation and increases apoptosis in fetal mouse brain. *J Nutr* **134**:162–166.
28. MI Berrocal-Zaragoza, JM Sequeira, MM Murphy, JD Fernandez-Ballart, SG Abdel Baki, PJ Bergold, EV Quadros. (2014) Folate

deficiency in rat pups during weaning causes learning and memory deficits. *Br J Nutr* **112**:1323–1332.
29. P Sable, A Kale, A Joshi, S Joshi. (2014) Maternal micronutrient imbalance alters gene expression of BDNF, NGF, TrkB and CREB in the offspring brain at an adult age. *Int J Dev Neurosci* **34**:24–32.
30. P Sable, K Randhir, A Kale, P Chavan-Gautam, S Joshi. (2015) Maternal micronutrients and brain global methylation patterns in the offspring. *Nutr Neurosci* **18**:30–36.
31. AP Meher, AA Joshi, SR Joshi. (2014) Maternal micronutrients, omega-3 fatty acids, and placental PPARgamma expression. *Appl Physiol Nutr Metab* **39**:793–800.
32. KD Dangat, AA Kale, SR Joshi. (2011) Maternal supplementation of omega 3 fatty acids to micronutrient-imbalanced diet improves lactation in rat. *Metabolism* **60**:1318–1324.
33. AP Meher, AA Joshi, SR Joshi. (2013) Preconceptional omega-3 fatty acid supplementation on a micronutrient-deficient diet improves the reproductive cycle in Wistar rats. *Reprod Fertil Dev* **25**:1085–1094.
34. MP Barnett, EN Bermingham, W Young, SA Bassett, JE Hesketh, A Maciel-Dominguez, WC McNabb, NC Roy. (2015) Low folate and selenium in the mouse maternal diet alters liver gene expression patterns in the offspring after weaning. *Nutrients* **7**:3370–3386.
35. CA Maloney, SM Hay, LE Young, KD Sinclair, WD Rees. (2011) A methyl-deficient diet fed to rat dams during the peri-conception period programs glucose homeostasis in adult male but not female offspring. *J Nutr* **141**:95–100.
36. CJ McNeil, SM Hay, GJ Rucklidge, MD Reid, GJ Duncan, WD Rees. (2009) Maternal diets deficient in folic acid and related methyl donors modify mechanisms associated with lipid metabolism in the fetal liver of the rat. *Br J Nutr* **102**:1445–1452.
37. CJ McNeil, SM Hay, GJ Rucklidge, MD Reid, GJ Duncan, CA Maloney, WD Rees. (2008) Disruption of lipid metabolism in the liver of the pregnant rat fed folate-deficient and methyl donor-deficient diets. *Br J Nutr* **99**:262–271.
38. CA Maloney, SM Hay, WD Rees. (2009) The effects of feeding rats diets deficient in folic acid and related methyl donors on the blood

pressure and glucose tolerance of the offspring. *Br J Nutr* **101**: 1333–1340.
39. CA Maloney, SM Hay, WD Rees. (2007) Folate deficiency during pregnancy impacts on methyl metabolism without affecting global DNA methylation in the rat fetus. *Br J Nutr* **97**:1090–1098.
40. KK Mejos, HW Kim, EM Lim, N Chang. (2013) Effects of parental folate deficiency on the folate content, global DNA methylation, and expressions of FRalpha, IGF-2 and IGF-1R in the postnatal rat liver. *Nutr Res Pract* **7**:281–286.
41. HW Kim, KN Kim, YJ Choi, N Chang. (2013) Effects of paternal folate deficiency on the expression of insulin-like growth factor-2 and global DNA methylation in the fetal brain. *Mol Nutr Food Res* **57**:671–676.
42. BG Swayne, A Kawata, NA Behan, A Williams, MG Wade, AJ Macfarlane, CL Yauk. (2012) Investigating the effects of dietary folic acid on sperm count, DNA damage and mutation in Balb/c mice. *Mutat Res* **737**:1–7.
43. S Ahmad, T Basak, K Anand Kumar, G Bhardwaj, A Lalitha, DK Yadav, GR Chandak, M Raghunath, S Sengupta. (2015) Maternal micronutrient deficiency leads to alteration in the kidney proteome in rat pups. *J Proteomics*.
44. Y Li, X Zhang, Y Sun, Q Feng, G Li, M Wang, X Cui, L Kang, Y Jiang. (2013) Folate deficiency during early-mid pregnancy affects the skeletal muscle transcriptome of piglets from a reciprocal cross. *PLoS One* **8**:e82616.
45. C Maestro-de-las-Casas, L Perez-Miguelsanz, Y Lopez-Gordillo, E Maldonado, T Partearroyo, G Varela-Moreiras, C Martinez-Alvarez. (2013) Maternal folic acid-deficient diet causes congenital malformations in the mouse eye. *Birth Defects Res A Clin Mol Teratol* **97**: 587–596.
46. KA Kumar, A Lalitha, D Pavithra, IJ Padmavathi, M Ganeshan, KR Rao, L Venu, N Balakrishna, NH Shanker, SU Reddy, GR Chandak, S Sengupta *et al.* (2013) Maternal dietary folate and/or vitamin B12 restrictions alter body composition (adiposity) and lipid metabolism in Wistar rat offspring. *J Nutr Biochem* **24**:25–31.
47. LH Allen, JL Rosado, JE Casterline, H Martinez, P Lopez, E Munoz, AK Black. (1995) Vitamin B-12 deficiency and malabsorption are

highly prevalent in rural Mexican communities. *Am J Clin Nutr* **62**:1013–1019.
48. KM Jones, M Ramirez-Zea, C Zuleta, LH Allen. (2007) Prevalent vitamin B-12 deficiency in twelve-month-old Guatemalan infants is predicted by maternal B-12 deficiency and infant diet. *J Nutr* **137**:1307–1313.
49. M Voutounou, CD Glen, YR Dubrova. (2012) The effects of methyl-donor deficiency on mutation induction and transgenerational instability in mice. *Mutat Res* **734**:1–4.
50. JL Buescher, LP Musselman, CA Wilson, T Lang, M Keleher, TJ Baranski, JG Duncan. (2013) Evidence for transgenerational metabolic programming in drosophila. *Dis Model Mech* **6**:1123–1132.
51. LM Matzkin, S Johnson, C Paight, TA Markow. (2013) Preadult parental diet affects offspring development and metabolism in Drosophila melanogaster. *PLoS ONE* **8**:e59530.
52. AO Joaquim, CP Coelho, PD Motta, FF Bondan, E Teodorov, MF Martins, TB Kirsten, RC Casarin, LV Bonamin, MM Bernardi. (2105) Transgenerational effects of a hypercaloric diet. *Reprod Fertil Dev*.
53. J Li, J Huang, JS Li, H Chen, K Huang, L Zheng. (2012) Accumulation of endoplasmic reticulum stress and lipogenesis in the liver through generational effects of high fat diets. *J Hepatol* **56**:900–907.
54. F Massiera, P Barbry, P Guesnet, A Joly, S Luquet, C Moreilhon-Brest, T Mohsen-Kanson, EZ Amri, G Ailhaud. (2010) A western-like fat diet is sufficient to induce a gradual enhancement in fat mass over generations. *J Lipid Res* **51**:2352–2361.
55. GC Burdge, SP Hoile, T Uller, NA Thomas, PD Gluckman, MA Hanson, KA Lillycrop. (2011) Progressive, transgenerational changes in offspring phenotype and epigenotype following nutritional transition. *PLoS ONE* **6**:e28282.
56. GA Dunn, TL Bale. (2009) Maternal high-fat diet promotes body length increases and insulin insensitivity in second-generation mice. *Endocrinology* **150**:4999–5009.
57. GA Dunn, TL Bale. (2011) Maternal high-fat diet effects on third-generation female body size via the paternal lineage. *Endocrinology* **152**:2228–2236.

58. T Fullston, NO Palmer, JA Owens, M Mitchell, HW Bakos, M Lane. (2012) Diet-induced paternal obesity in the absence of diabetes diminishes the reproductive health of two subsequent generations of mice. *Human Reproduction* **27**:1391–1400.
59. T Fullston, EM Ohlsson Teague, NO Palmer, MJ DeBlasio, M Mitchell, M Corbett, CG Print, JA Owens, M Lane. (2013) Paternal obesity initiates metabolic disturbances in two generations of mice with incomplete penetrance to the F2 generation and alters the transcriptional profile of testis and sperm microRNA content. *FASEB J* **27**:4226–4243.
60. NO McPherson, T Fullston, HW Bakos, BP Setchell, M Lane. (2014) Obese father's metabolic state, adiposity, and reproductive capacity indicate son's reproductive health. *Fertil Steril* **101**:865–873.
61. RC Painter, C Osmond, P Gluckman, M Hanson, DI Phillips, TJ Roseboom TJ. (2008) Transgenerational effects of prenatal exposure to the Dutch famine on neonatal adiposity and health in later life. *BJOG* **115**:1243–1249.
62. MV Veenendaal, RC Painter, SR de Rooij, PM Bossuyt, JA van der Post, PD Gluckman, MA Hanson, TJ Roseboom. (2013) Transgenerational effects of prenatal exposure to the 1944–45 Dutch famine. *BJOG* **120**:548–553.
63. G Kaati, LO Bygren, S Edvinsson. (2002) Cardiovascular and diabetes mortality determined by nutrition during parents' and grandparents' slow growth period. *Eur J Hum Genet* **10**:682–688.
64. ME Pembrey, LO Bygren, G Kaati, S Edvinsson, K Northstone, M Sjostrom, J Golding. (2006) Sex-specific, male-line transgenerational responses in humans. *European Journal of Human Genetics* **14**:159–166.
65. LO Bygren, G Kaati, S Edvinsson. (2001) Longevity determined by paternal ancestors' nutrition during their slow growth period. *Acta Biotheor* **49**:53–59.
66. C Bertram, O Khan, S Ohri, DI Phillips, SG Matthews, MA Hanson. (2008) Transgenerational effects of prenatal nutrient restriction on cardiovascular and hypothalamic-pituitary-adrenal function. *J Physiol* **586**:2217–2229.

67. DC Benyshek, CS Johnston, JF Martin. (2006) Glucose metabolism is altered in the adequately-nourished grand-offspring (F3 generation) of rats malnourished during gestation and perinatal life. *Diabetologia* **49**:1117–1119.
68. O Rechavi, L Houri-Ze'evi, S Anava, WS Goh, SY Kerk, GJ Hannon, O Hobert. (2014) Starvation-induced transgenerational inheritance of small RNAs in C. elegans. *Cell* **158**:277–287.
69. H Pijl. (2011) Obesity: evolution of a symptom of affluence. *Neth J Med* **69**:159–166.
70. MJ Edwards. (2012) Genetic selection of embryos that later develop the metabolic syndrome. *Med Hypotheses* **78**:621–625.
71. SP Hoile, KA Lillycrop, NA Thomas, HA Hanson, GC Burdge. (2011) Dietary protein restriction during F0 pregnancy in rats induces transgenerational changes in the hepatic transcriptome in female offspring. *PLoS ONE* **6**:e21668.
72. E Zambrano, PM Martinez-Samayoa, CJ Bautista, M Deas, L Guillen, GL Rodriguez-Gonzalez, C Guzman, F Larrea, PW Nathanielsz. (2005) Sex differences in transgenerational alterations of growth and metabolism in progeny (F2) of female offspring (F1) of rats fed a low protein diet during pregnancy and lactation. *J Physiol* **566**:225–236.
73. E Zambrano, CJ Bautista, M Deas, PM Martinez-Samayoa, M Gonzalez-Zamorano, H Ledesma, J Morales, F Larrea, PW Nathanielsz. (2006) A low maternal protein diet during pregnancy and lactation has sex- and window of exposure-specific effects on offspring growth and food intake, glucose metabolism and serum leptin in the rat. *J Physiol* **571**:221–230.
74. ED Frantz, MB Aguila, R Pinheiro-Mulder Ada, CA Mandarim-de-Lacerda. (2011) Transgenerational endocrine pancreatic adaptation in mice from maternal protein restriction in utero. *Mech Ageing Dev* **132**:110–116.
75. BF Ponzio, MH Carvalho, ZB Fortes, M do Carmo Franco. (2012) Implications of maternal nutrient restriction in transgenerational programming of hypertension and endothelial dysfunction across F1-F3 offspring. *Life Sciences* **90**:571–577.
76. FA Nascimento, TC Ceciliano, MB Aguila, CA Mandarim-de-Lacerda. (2013) Transgenerational effects on the liver and pancreas resulting

from maternal vitamin D restriction in mice. *J Nutr Sci Vitaminol* (Tokyo) **59**:367–374.
77. DA Fernandes de Abreu, V Landel, AG Barnett, J McGrath, D Eyles, F Feron. (2012) Prenatal vitamin D deficiency induces an early and more severe experimental autoimmune encephalomyelitis in the second generation. *Int J Mol Sci* **13**:10911–10919.
78. T Scully. (2012) Diabetes in numbers. *Nature* **485**:S2–3.
79. W Reik, W Dean, J Walter. (2001) Epigenetic reprogramming in mammalian development. *Science* **293**:1089–1093.
80. W Reik, F Santos, W Dean. (2003) Mammalian epigenomics: reprogramming the genome for development and therapy. Theriogenology **59**:21–32.
81. MB Vandegehuchte, CR Janssen. (2013) Epigenetics in an ecotoxicological context. *Mutation Research*.
82. SG Gu, J Pak, S Guang, JM Maniar, S Kennedy, A Fire. (2012) Amplification of siRNA in Caenorhabditis elegans generates a transgenerational sequence-targeted histone H3 lysine 9 methylation footprint. *Nature Genetics* **44**:157–164.
83. A Ashe, A Sapetschnig, EM Weick, J Mitchell, MP Bagijn, AC Cording, AL Doebley, LD Goldstein, NJ Lehrbach, J Le Pen, G Pintacuda, A Sakaguchi *et al.* (2012) piRNAs can trigger a multigenerational epigenetic memory in the germline of C. elegans. *Cell* **150**:88–99.
84. CS Rosenfeld. (2014) Animal models of transgenerational epigenetic effects, in *Transgenerational Epigenetics,* T Tollefsbol, Editor. London, UK: Elsevier Publications, 123–145.

Epigenetic Biomarkers and Global Health

CHAPTER 8

Paula Costello* and Mark Hanson*,†

*Institute of Developmental Sciences, University of Southampton, IDS Building (MP 887), Southampton General Hospital, Tremona Road, Southampton, SO16 6YD, UK

†NIHR Southampton Biomedical Research Centre, Southampton Centre for Biomedical Research (MP218), University of Southampton and University Hospital, Southampton NHS Trust, Tremona Road, Southampton, SO16 6YD, UK

The Global Burden of Non-communicable Diseases

The prevalence of non-communicable diseases (NCDs) is increasing globally at an alarming rate: the World Health Organization predicts that by 2030 they will be the leading cause of death and disability in every region of the world.[1] NCDs were once thought of as diseases of the affluent but in fact 80% of the deaths from NCDs are now occurring in low- and middle-income countries. Furthermore, nearly 30% of NCD-related deaths in low-income countries occur under the age of 60 compared to only 13% in high-income countries. There is therefore an urgent need to address the NCD epidemic and to implement effective interventions. Despite the scale of the challenge posed by NCDs, the issue was not identified in the Millennium Development Goals and it was not until September 2011 that the United Nations called a High Level summit on the prevention and control of NCDs, the second time only that heads of state have convened around a health issue. The political declaration arising from the summit[2] highlighted the urgency surrounding NCDs and recognized the importance of taking a developmental perspective to their prevention. NCDs are chronic diseases of long duration and generally slow progression. The four main disease groups of NCDs are diabetes, cancer, cardiovascular disease and respiratory disease. NCDs are caused to a large extent by four behavioral risk factors: tobacco use, poor nutrition, physical inactivity and harmful use of alcohol.[3] These behaviors in turn lead to four key metabolic/physiological changes that increase the risk of NCDs: raised blood pressure, obesity, hyperglycemia and hyperlipidemia. The risk of NCDs increases over the life course, starting in early development[4] especially with overweight or obesity.[5] The prevalence of childhood obesity is increasing rapidly and this not only has health implications for the child but is also linked to an increased risk of adult obesity and its related adverse metabolic consequences in later life.[6] Diet quality in early childhood is a determinant of obesity risk, with low quality diets (characterized by high consumption of energy-dense foods and low consumption of fruits, vegetables and fish) linked to obesity risk.[7] Dietary behaviors established in early childhood have also been shown to track into adulthood.[8] A further cause for action is that risk factors for childhood obesity are modifiable: they include maternal smoking during pregnancy, maternal obesity before pregnancy and excessive gestational weight gain, low maternal vitamin D status, short breastfeeding duration, and inadequate infant sleep duration.[9] This is a part of the growing evidence that the early life environment can play an important role in influencing the risk of developing NCDs in later life.[10] To date, fixed genomic variations such as single nucleotide polymorphisms and copy number variations have been shown to account for only a small proportion of the variation in NCD risk.[11] It is therefore thought that the developmental environment can influence later phenotype though the altered epigenetic regulation of genes leading to persistent phenotypic changes and an altered risk of NCDs in later life.

What is an Epigenetic Biomarker?

Biomarkers are biological measures of a biological state and in medicine a biomarker is a measurable indicator of the severity or presence of some disease state. However, more generally, a biomarker is anything that can be used as an indicator of the physical state of an organism and therefore used as a measure of health as well as disease risk. Biomarkers can include physiological indicators such as blood pressure or molecular markers such as liver enzymes and they can serve as a predictor of future health or disease. They can provide useful information even when a detailed understanding of the underlying mechanism is lacking.

Many mechanisms of epigenetic regulation have been discovered in recent years and the study of epigenetic biomarkers has become a rapidly advancing field. Epigenetic processes including DNA methylation and histone modifications regulate gene expression by modulating the packaging of the DNA without a change in genomic sequence[12] and these changes can be maintained over multiple divisions in somatic cells. DNA methylation can act directly to block binding of transcription factors to the DNA or by recruiting other repressive factors which in turn mediate local chromatin changes.[13] Histone modifications such as acetylation, methylation, ubiquitination and phosphorylation can directly affect chromatin structure and therefore the accessibility of the underlying genomic sequence, they also provide binding sites for proteins involved in gene regulation. Other mechanisms include the non-coding RNAs (ncRNA), RNAs which are not transcribed but can induce mRNA degradation or translational repression and when targeted to the promoter region of a gene, can induce both DNA methylation and repressive histone modifications.[14] They may also act at enhancer or repressor sites. Environmental and lifestyle-related influences such as nutrition and exposure to stress can induce epigenetic alterations and so the epigenome can therefore be regarded as a molecular record of previous life events. For example, monozygotic twins have been shown to be epigenetically similar at birth but their epigenomes diverge with age and at a rate that is increased if they do not share a common environment.[15] The developmental environment can alter later phenotype through epigenetic mechanisms and the plasticity of the developmental epigenome is thought to have adaptive value since it allows the fetus to predict and prepare for the environment likely to be experienced later.[16] Many of the targets of epigenetic processes during development are transcription factor binding sites, and their effects will not be manifested until the appropriate transcription factor is present. For this reason, epigenetic effects of the developmental environment will not necessarily be manifested until later in the life course. Epigenetic biomarkers therefore have the potential to be used both as a measure of the individual's otherwise hidden exposure to signals during development and as a predictor of the individual's later disease risk, enabling early intervention strategies to improve both early development and later health.

Global Health Challenges and Epigenetics

New and ongoing research is uncovering the role of epigenetics in setting the risk of a variety of NCDs. Cancer has been analyzed in the most detail and it is well-known that the epigenetic silencing of tumor suppressor genes is a frequent event in multiple cancers.[17] Epigenetic deregulation in a cell may also determine a cancer fate long before it is visually identifiable as a tumor cell.[18] Cancer epigenetic biomarkers could therefore be used as a cancer risk assessment long before the clinical symptoms appear. Epigenetic mechanisms have also been shown to contribute to the development of autoimmune diseases such as type I diabetes,[19] in the inflammation associated with cardiovascular disease,[20] hypertension[21] and respiratory disease such as asthma and COPD.[22]

There are now many studies focused on identifying the link between an adverse *in utero* environment and epigenetic modifications that take place in the developing fetus which may alter susceptibility to adult-onset chronic disease. Since the epigenome is most susceptible to environmental factors in early life this offers the potential opportunity for intervention to reverse marks associated with disease risk. A well-known example of maternal nutrition influencing offspring phenotype via epigenetic mechanisms is the agouti mouse model, where coat color is determined by the methylation status of the 5′ end of the *Agouti* gene and methylation of the gene is altered by mother's intake of dietary methyl donors and co-factors.[23,24] Studies in other animal models have also shown that maternal diet can establish persistent metabolic changes in the offspring through epigenetic changes in key metabolic genes. For example, a protein restricted diet in pregnant rats induced hypomethylation of the glucocorticoid receptor (GR) and peroxisome proliferator activated receptor alpha (PPARα) in the livers of juvenile and adult offspring and this was accompanied by increased GR and PPARα gene expression and a persistent change in the metabolic processes, namely gluconeogenesis and fatty acid beta oxidation, that these nuclear receptors control.[25,26]

Given the growing concern over the widely consumed energy-dense Western diet, many studies are focusing on the effects of a maternal high fat diet. In rats, maternal high fat feeding during pregnancy leads to reduced expression of FASD2, the rate-limiting enzyme in polyunsaturated fatty acid synthesis in the liver of the offspring and this is accompanied by altered methylation of CpGs in the promoter of the gene.[27] Maternal obesity and diabetes in mice has also been shown to induce widespread changes in DNA methylation in the offspring's liver.[28] The period of epigenetic plasticity may also extend into post-natal life. Overfeeding in rat pups was shown to induce hypermethylation of two CpG nucleotides within the proopiomelanocortin (POMC) promoter, a gene which plays a key role in appetite control, and this hypermethylation prevented upregulation of POMC expression when plasma levels of both leptin and insulin were high.[29]

DNA methylation is the most stable epigenetic mark and the easiest to measure so it is the most widely studied but there are also studies showing how persistent changes can be induced in both histone modifications and ncRNAs. In an obese pregnant monkey model the offspring were also obese and were found to have site-specific alterations in fetal hepatic H3 acetylation.[30] Changes in miRNA (miR) expression have also been shown in the liver and skeletal muscle of sheep following a protein restriction during pregnancy.[31,32]

Practicalities of Epigenetic Biomarkers in Global Health

In epigenetic biomarker discovery it will often be impracticable to study the diseased or appropriate tissues and so more accessible surrogate tissues such as peripheral blood must sometimes be used instead. Although DNA methylation patterns are thought to be tissue specific, a number of studies have also shown inter-tissue methylation correlations. Murphy et al. (2012)[33] showed that the methylation across differentially methylated regions for the imprinted genes *H19*, *MEST* and *PEG10* did not significantly differ across a wide range of human tissues. The methylation levels of a number of non-imprinted genes measured in blood were also equivalent in buccal cells despite these cell types originating from different germ layers (mesoderm and ectoderm respectively).[34] Methylation changes induced by maternal diet were also found to be similar in the umbilical cord and liver.[35] Even if the epigenetic marks in the proxy tissue do not represent the methylation of the functional tissue, if they can repeatedly predict disease or risk, they could still be classified as a biomarker and used accordingly.

Peripheral blood is an obvious choice for a proxy tissue when sampling large populations and would allow samples to be taken repeatedly throughout a longitudinal study, plus DNA from existing biobanks are generally extracted from whole blood and so could be used to perform epigenetic association studies. DNA methylation has been found to be stable over time in both blood and buccal cell samples and the methylation measured in blood correlated with that in the buccal cells.[34] However, peripheral blood is a heterogeneous mix of blood cell populations which may be a potential confounder since different white blood cells have specific DNA methylation patterns.[36] Inter-individual variation may occur whereby two individuals differ in their cellular heterogeneity and therefore in their measured methylation levels at specific loci, but in the single cell population the methylation may be the same. Intra-individual variation may also occur in longitudinal studies due to changes in the relative proportions of the different cell types. Taking a full blood count in order to determine the relative proportions of each cell type and correcting for this in the analysis would easily avoid these disadvantages. miRs are now widely measured in many studies as they are associated with some disease states[37] and are relatively stable in blood samples and even dried blood spots.[38]

Recent advances in high-throughput sequencing have enabled epigenetic biomarker discovery on a genome-wide scale. The increased coverage of the new platforms is likely to uncover many new disease-related epigenetic modifications which are located outside well-known candidate regions such as CpG islands and gene promoters. The Roadmap Epigenomics Project is producing reference epigenomes for 127 tissue and cell types from healthy donors and people with diseases.[39] The tissues have been profiled for histone modification patterns, DNA accessibility, DNA methylation and RNA expression, and they will become an invaluable tool to efficiently compare epigenomes in different cells.

The Application of Epigenetic Biomarkers

Unlike communicable diseases, in which exposure to a pathogenic organism, usually through direct or indirect contact with an infected individual, is an essential step for contracting the disease, for NCDs *everyone* is at risk. This is because the processes involved in the aetiology of NCDs involve mechanisms that are fundamental to human biology, namely development and aging. During development, the embryo, fetus and infant detect aspects of their environment, especially those associated with the nutrition, body composition and behavior of parents, and modify their own phenotypes accordingly to be best prepared for life in that environment. The degree of "match" or "mismatch"[40] between the phenotype and the environment is a critical aspect of the risk of later NCDs, which increases throughout the life course. The life course concept (Figure 1) explains why the risk of NCDs is not solely due to either inherited fixed genetic factors or adult lifestyle, although both play a part in setting risk. In this context, "risk" has to be viewed not just as an absolute level which has been reached over the life course to date, but as the degree of responsiveness to new challenges, such as those imposed by adult lifestyle factors including diet, physical activity, stress levels etc.

In individuals with poor responses to such challenges, the detrimental effects on subsequent health will be greater. As shown in

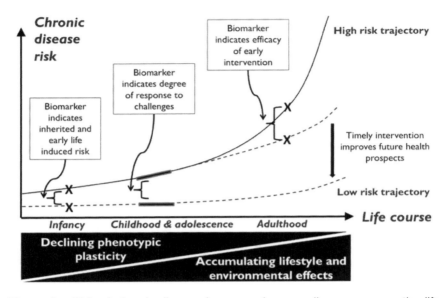

Figure 1. Risk of chronic disease increases in a non-linear way over the life course as a result of declining phenotypic plasticity and accumulative damage from lifestyle-imposed and other environmental challenges. Heritable genetic, epigenetic and environmental factors in early life can also set an individual on a higher risk trajectory. Screening for biomarkers at infancy could identify such individuals and allow timely intervention in childhood and adolescence to lower their risk and improve future health prospects. Such biomarkers could also be used at later points in the life course to indicate the degree of response to challenges and the efficacy of early interventions.

Figure 1, these considerations apply to differences between individuals in risk inherited from the previous generation and operating via: developmental processes; the peak function in an organ or system obtained; the rate of decline with aging, including any contextual measures available to reduce this decline; the overall level of health care, which varies substantially between low and high income countries; and to external risk factors such as exposure to endocrine disruptor chemicals.[41]

These concepts underscore the need for biomarkers of risk to be applied at various points along the life course, especially in early development, adulthood and aging. These biomarkers should ideally not only be a measure of absolute risk but also measure the responsiveness to challenges. The first type of biomarker would be a measure of pathophysiological process. The second would indicate that the biomarker is informative if it is itself a part of the physiological pathway of compensatory responses. Epigenetic biomarkers may form part of either of these components: pathophysiological changes that have been shown to be relevant in diseases such as cancer, atherosclerosis, and diabetes (see above) or a second class shown to be involved in regulatory processes such as fat metabolism,[42,43] fat deposition,[44] or mitochondrial function.[45,46] As noted above, epigenetic biomarkers are particularly important in the early developmental effects of environment on later disease risk.

The life course concept of responsiveness and risk is particularly important in young people. They are a section of the population who very often do not receive the foremost attention of primary health care providers and who indeed often believe themselves healthy. Moreover, this is a section of the population who discount current versus future health, especially among those with lower socioeconomic status or educational attainment. Such young people may indeed be relatively healthy in terms of their ongoing pathology, but this is certainly not true for some of them in terms of their risk trajectories. Biomarkers may therefore be helpful in identifying those with an inherently higher risk trajectory in order to help reduce that risk. Biomarkers at this time in the life course can give valuable retrospective information on past exposure to risk and also prospective information on later risk of NCDs. This has been shown for example by the effects of maternal nutrition in pregnancy on epigenetic biomarkers at birth in the infant and on later adiposity at 6–9 years in the child.[44]

The focus on younger members of the population is particularly important as the risk of NCDs induced by poor diet, obesity or unhealthy behaviors has been shown to be passed across generations by processes which include epigenetic mechanisms.[47] This is a novel concept, by which epigenetic processes can be induced in the germ cells during their formation,[48] but also in some somatic tissues which can directly or indirectly mediate the passage of epigenetic processes to the next generation.[49] This is one aspect of the transmission of health capital from one generation to the next referred to above. Of particular note is that this pathway has now been shown to operate via both male and female lineages, the former involving mechanisms including small non-coding RNAs in sperm which are influenced by paternal behavioral factors[50] and may affect the development of the early embryo. Thus, whilst epigenetic marks in the ovum may indicate grand-maternal as well as maternal life course influences extending back over many years, those in the sperm may indicate much more recent influences.

Epigenetic biomarkers may be an invaluable tool in stratifying the population for risk, or in identifying sections of the population at particular risk of NCDs. They may therefore be used to target interventions accordingly, or to permit the application of specific interventions to particular population groups. These considerations are especially important in devising an intervention strategy which may reduce social inequalities in health, as well as addressing specific risk in particular sections of the population.[51] Moreover, although some epigenetic marks are induced during early development, the period of developmental plasticity may extend well into post-natal life. This offers the possibility that some epigenetic changes, e.g. those induced pre-natally by poor maternal diet, may be reversed by appropriate post-natal interventions. In addition, recent evidence shows that adolescence may constitute another period of phenotypic plasticity.[52,53] For all these reasons, epigenetic biomarkers may be useful not only in measuring risk of NCDs but also in monitoring the efficacy of interventions.

Lastly, as we learn more of the mechanistic basis for the epigenetic changes, biomarkers may suggest novel avenues for interventions, including new drug targets. This aspect of their use is still in the very early stages of development, since current drugs which interfere with epigenetic processes have widespread and therefore undesirable effects: they can be useful therapeutically in conditions such as forms of cancer where epigenetic changes are involved in the pathology, but are not as yet feasible for use in preventative interventions to reduce risk earlier in the life course.

Raising Awareness at a National Level

One of the major challenges in reducing NCD risk at the population level concerns raising awareness of the problem in a non-judgmental way which promotes self-efficacy in at-risk individuals and groups. This is particularly true of lower socioeconomic status groups and those with lower educational attainment, and of migrants or displaced population groups. All of these groups have lower access to primary health care. There is unfortunately a widespread belief that NCDs are due to inherited genetic risk and that the only solution lies in medical treatment such as drug therapy, bariatric surgery etc. This may be because such an apparently fatalistic view abrogates personal responsibility for preventative action. This is made worse by the view among some authorities, from public health departments to educators and social workers, that risk is entirely a matter of personal choice, increasing a sense of blame in those affected, e.g. those who are overweight or obese.[54] Personal choice is clearly not applicable to young children, in whom risk of later NCDs can be induced; nor should it necessarily be applied to older individuals, in the light of recent findings that early life factors can influence the development of appetite control, food preferences or taste[55] which are then perhaps beyond an individual's control later.

These issues might be addressed in part by the wider dissemination of information about epigenetic biomarkers. They may help to promote wider understanding of the life course concept of NCD risk; to make it clear that all members of the population, including those younger people who believe that they are completely healthy, have some degree of risk; and to counter the fatalistic view that individuals can do little to reduce their risk. Wider public discussion of the role of lifestyle factors in inducing changes in biomarkers and of their role in NCD risk across the life course may be helpful in raising awareness of the NCD issue, in particular because the possibility exists that lifestyle behaviors can reduce or even reverse risk, and it may be possible to demonstrate such effects on biomarkers in the short term before other phenotypic markers such as adiposity or blood pressure have changed. This may empower individuals to make such lifestyle changes. This necessitates increasing health literacy in susceptible groups such as adolescents[56] but the complexities of translating such increased awareness into behavior change should not be underestimated.[57]

Public health policy at the level of individual societies and countries should incorporate knowledge of epigenetic biomarkers. Information about these might be incorporated into educational curricula for schools, for initiatives to promote health literacy and behavior change through learning outside the classroom and to wider campaigns for health promotion.

Conclusion: Implications of Epigenetic Biomarkers for Global Health

The importance of epigenetic biomarkers discussed above has wider implications for global health issues. Many of the challenges which form the basis for the Millennium Development Goals, and which are even more apparent in the Sustainable Development Goals, can be seen to have potential impact on epigenetic processes across the life course and across generations. They may therefore be captured by epigenetic biomarkers which could be used to monitor progress towards these goals. Many of the factors associated with epigenetic changes are associated with poverty, including poor diets, exposure to environmental toxicants and pollution and high levels of stress. Poor educational and socioeconomic status are associated with social inequalities in health which may be manifest through epigenetic processes. Increasing urbanization and adoption of a Westernized lifestyle are producing dramatic effects on NCD incidence, partly through epigenetic processes. From an economic point of view, the influence of global multinational companies on food choices and lifestyle factors such as sedentary activities are similarly having dramatic effects. Lastly, in indirect ways, climate change is influencing patterns of crop and food production, air quality and demography, all of which affect human populations in multiple ways and risk of NCDs. It may be that epigenetic biomarkers, as sensitive indicators of the ways in which many of these factors influence human health, could serve as part of a package of measures of sustainable development — not only like the canary in the coal mine signaling higher levels of risk, but as indicators of effective interventions.

Acknowledgement

MAH is supported by the British Heart Foundation.

References

1. World Health Organization. (2015) Non-communicable diseases fact sheet, http://www.who.int/mediacentre/factsheets/fs355/en/.
2. U.N. General Assembly. (2012) U.N.G. Resolution adopted by the General Assembly 66/2. Political Declaration on the Prevention and Control of Non-communicable Diseases.
3. A Alwan. (2011) *Global Status Report On Non-communicable Diseases 2010*. World Health Organization.
4. KM Godfrey, PD Gluckman, MA Hanson. (2010) Developmental origins of metabolic disease: life course and intergenerational perspectives. *Trends Endocrinol Metab* **21**:199–205.
5. Global Burden of Metabolic Risk Factors for Chronic Diseases Collaboration. (2014) Metabolic mediators of the effects of body-mass index, overweight, and obesity on coronary heart disease and stroke: a pooled analysis of 97 prospective cohorts with 1.8 million participants. *Lancet* **383**:970–983.
6. JJ Reilly, J Kelly. (2011) Long-term impact of overweight and obesity in childhood and adolescence on morbidity and premature mortality in adulthood: systematic review. *Int J Obes* **35**:891–898.
7. H Okubo *et al.* (2015) Diet quality across early childhood and adiposity at 6 years: the Southampton Women/'s Survey. *Int J Obes*.

8. V Mikkilä, L Räsänen, OT Raitakari, P Pietinen, J Viikari. (2005) Consistent dietary patterns identified from childhood to adulthood: the cardiovascular risk in young Finns study. *British Journal of Nutrition* **93**:923–931.
9. MW Gillman et al. (2008) Developmental origins of childhood overweight: potential public health impact. *Obesity (Silver Spring, Md.)* **16**:1651–1656.
10. KM Godfrey, HM Inskip, MA Hanson. (2011) The long term effects of prenatal development on growth and metabolism. *Seminars in Reproductive Medicine* **29**:257–265.
11. AP Morris et al. (2012) Large-scale association analysis provides insights into the genetic architecture and pathophysiology of type 2 diabetes. *Nature Genetics* **44**:981–990.
12. A Bird, D Macleod. (2004) Reading the DNA methylation signal. *Cold Spring Harb Symp Quant Biol* **69**:113–118.
13. F Fuks et al. (2003) The methyl-CpG-binding protein MeCP2 links DNA methylation to histone methylation. *Journal of Biological Chemistry* **278**:4035–4040.
14. PG Hawkins, S Santoso, C Adams, V Anest, KV Morris. (2009) Promoter targeted small RNAs induce long-term transcriptional gene silencing in human cells. *Nucleic Acids Research.*
15. MF Fraga et al. (2005) Epigenetic differences arise during the lifetime of monozygotic twins. *Proc Natl Acad Sci U S A* **102**:10604–10609.
16. PD Gluckman, MA Hanson, AS Beedle, HG Spencer. (2008) Predictive adaptive responses in perspective. *Trends Endocrinol Metab* **19**:109–110.
17. AP Feinberg, B Tycko. (2004) The history of cancer epigenetics. *Nat Rev Cancer* **4**:143–153.
18. AP Feinberg, R Ohlsson, S Henikoff. (2006) The epigenetic progenitor origin of human cancer. *Nat Rev Genet* **7**:21–33.
19. MN Dang, R Buzzetti, P Pozzilli. (2013) Epigenetics in autoimmune diseases with focus on type 1 diabetes. *Diabetes/Metabolism Research and Reviews* **29**:8–18.
20. P Stenvinkel et al. (2007) Impact of inflammation on epigenetic DNA methylation — a novel risk factor for cardiovascular disease? *Journal of Internal Medicine* **261**:488–499.

21. L Raftopoulos *et al.* (2015) Epigenetics, the missing link in hypertension. *Life Sciences* **129**:22–26.
22. M Kabesch, IM Adcock. (2012) Epigenetics in asthma and COPD. *Biochimie* **94**:2231–2241.
23. HD Morgan, HG Sutherland, DI Martin, E Whitelaw. (1999) Epigenetic inheritance at the agouti locus in the mouse. *Nat Genet* **23**:314–318.
24. RA Waterland, RL Jirtle. (2003) Transposable elements: targets for early nutritional effects on epigenetic gene regulation. *Mol Cell Biol* **23**:5293–300.
25. KA Lillycrop, ES Phillips, AA Jackson, MA Hanson, GC Burdge. (2005) Dietary protein restriction of pregnant rats induces and folic acid supplementation prevents epigenetic modification of hepatic gene expression in the offspring. *Journal of Nutrition* **135**:1382–1386.
26. KA Lillycrop *et al.* (2008) Feeding pregnant rats a protein-restricted diet persistently alters the methylation of specific cytosines in the hepatic PPARalpha promoter of the offspring. *British Journal of Nutrition* **100**:278–282.
27. SP Hoile *et al.* (2013) Maternal fat intake in rats alters 20:4n-6 and 22:6n-3 status and the epigenetic regulation of Fads2 in offspring liver. *J Nutr Biochem.*
28. CCY Li *et al.* (2013) Maternal obesity and diabetes induces latent metabolic defects and widespread epigenetic changes in isogenic mice. *Epigenetics* **8**:602–611.
29. A Plagemann *et al.* (2009) Hypothalamic proopiomelanocortin promoter methylation becomes altered by early overfeeding: an epigenetic model of obesity and the metabolic syndrome. *J Physiol* **587**:4963–4976.
30. KM Aagaard-Tillery *et al.* (2008) Developmental origins of disease and determinants of chromatin structure: maternal diet modifies the primate fetal epigenome. *Journal of Molecular Endocrinology* **41**: 91–102.
31. S Lie *et al.* (2014) Periconceptional undernutrition programs changes in insulin-signaling molecules and microRNAs in skeletal muscle in singleton and twin fetal sheep. *Biology of Reproduction* **90**(5):1–10.
32. S Lie *et al.* (2014) Impact of embryo number and maternal undernutrition around the time of conception on insulin signaling and gluconeogenic

factors and microRNAs in the liver of fetal sheep. *Am J Physiol Endocrinol Metab* **306**:E1013–E1024.
33. SK Murphy, Z Huang, C Hoyo. (2012) Differentially methylated regions of imprinted genes in prenatal, perinatal and postnatal human tissues. *PLoS One* **7**:e40924.
34. RP Talens *et al.* (2010) Variation, patterns, and temporal stability of DNA methylation: considerations for epigenetic epidemiology. *The FASEB Journal* **24**:3135–3144.
35. GC Burdge, MA Hanson, JL Slater-Jefferies, KA Lillycrop. (2007) Epigenetic regulation of transcription: a mechanism for inducing variations in phenotype (fetal programming) by differences in nutrition during early life? *Br J Nutr* **97**:1036–1046.
36. BT Adalsteinsson *et al.* (2012) Heterogeneity in white blood cells has potential to confound DNA methylation measurements. *PLoS ONE* **7**:e46705.
37. T Thum, M Mayr. (2012) Review focus on the role of microRNA in cardiovascular biology and disease. *Cardiovasc Res* **93**:543–544.
38. Z Sergueeva, H Collins, ML Parrish, M McWhorter, S Dow. (2012) *Novel Tissue Types for the Development of Genomic Biomarkers.* INTECH Open Access Publisher.
39. CE Romanoski, CK Glass, HG Stunnenberg, L Wilson, G Almouzni. (2015) Epigenomics: roadmap for regulation. *Nature* **518**:314–316.
40. P Gluckman, M Hanson. (2006) *Mismatch: Why our world no longer fits our bodies.* Oxford University Press.
41. TT Schug *et al.* (2013) PPTOX III: environmental stressors in the developmental origins of disease — evidence and mechanisms. *Toxicol Sci* **131**:343–350.
42. CG Burdge, KA Lillycrop. (2014) Fatty acids and epigenetics. *Curr Opin Clin Nutr Metab Care* **17**:156–161.
43. H Masuyama, Y Hiramatsu. (2012) Effects of a high-fat diet exposure in utero on the metabolic syndrome-like phenomenon in mouse offspring through epigenetic changes in adipocytokine gene expression. *Endocrinology* **153**:2823–2830.
44. KM Godfrey *et al.* (2011). Epigenetic gene promoter methylation at birth is associated with child's later adiposity. *Diabetes* **60**:1528–1534.

45. PF Chinnery, HR Elliott, G Hudson, DC Samuels, CL Relton. (2012) Epigenetics, epidemiology and mitochondrial DNA diseases. *Int J Epidemiol* **41**:177–187.
46. CJ Pirola *et al.* (2013) Epigenetic modification of liver mitochondrial DNA is associated with histological severity of nonalcoholic fatty liver disease. *Gut* **62**:1356–1363.
47. MA Hanson, PD Gluckman. (2014) Early developmental conditioning of later health and disease: physiology or pathophysiology? *Physiol Rev* **94**:1027–1076.
48. MA Surani. (2015) Human germline: a new research frontier. *Stem Cell Reports* **4**:955–960.
49. MASMK Hanson. (In Press) Developmental origins of epigenetic transgenerational inheritance, in *Enviromental Epigenetics*.
50. EL Marczylo, AA Amoako, JC Konje, TW Gant, TH Marczylo. (2012) Smoking induces differential miRNA expression in human spermatozoa: a potential transgenerational epigenetic concern? *Epigenetics* **7**:432–439.
51. J Benach, D Malmusi, Y Yasui, JM Martinez. (2013) A new typology of policies to tackle health inequalities and scenarios of impact based on Rose's population approach. *J Epidemiol Community Health* **67**:286–291.
52. KC Kadosh, DE Linden, JY Lau. (2013) Plasticity during childhood and adolescence: innovative approaches to investigating neurocognitive development. *Developmental science* **16**:574–583.
53. SM Sawyer *et al.* (2012) Adolescence: a foundation for future health. *Lancet* **379**:1630–1640.
54. JD Latner, RM Puhl, JM Murakami, KS O'Brien. (2014) Food addiction as a causal model of obesity. Effects on stigma, blame, and perceived psychopathology. *Appetite* **77**:77–82.
55. JA Mennella. (2014) Ontogeny of taste preferences: basic biology and implications for health. *Am J Clin Nutr* **99**:704S–11S.
56. M Grace *et al.* (2012) Developing teenagers' views on their health and the health of their future children. *Health Education* **112**:543–559.
57. S Michie, MM van Stralen, R West. (2011) The behavior change wheel: a new method for characterising and designing behavior change interventions. *Implement Sci* **6**:42.

Nutrition, Epigenetics and Health: Evolutionary Perspectives

Sinead English*,† and Tobias Uller*,‡

*Edward Grey Institute,
Department of Zoology, University of Oxford,
The Tinbergen Building, South Parks Road,
Oxford, OX1 3PS, UK
†Department of Zoology, University of Cambridge,
Downing St, Cambridge, CB2 3EJ, UK
‡Department of Biology, Lund University,
Sölvegatan 35, 223 62 Lund, Sweden

Introduction

All animals need to find food and attain a balanced diet for successful growth, survival and reproduction. As such, ecologists and evolutionary biologists have long appreciated that nutrition is a powerful cause of differences between individuals, populations and species.[1] The amount of fairy shrimp ingested by spadefoot toad tadpoles during development, for example, triggers a dramatic switch between developing as a small omnivore or a large cannibalistic carnivore.[2] Consistent dietary differences between populations generate selection that can contribute to genetic divergence and local adaptation, and features that separate species often reflect their diets. Adaptation to diet may thus be a driver of speciation. For example, changes in beak morphology of Darwin's ground finches, which evolved in response to different feeding substrates between islands of the Galapagos, have resulted in changes in male song that facilitated reproductive isolation among subpopulations.[3]

Nutrition has also played an important role in human evolution. Like all animals, human adaptations partly reflect what we eat. This includes our teeth and digestive system, as well as our social behaviors, such as parental care. Nutrition can also be a strong cause of selection in humans, as evident from mass mortality due to starvation and nutritional deficits around the world. The frequency of such events before the formation of sedentary societies is largely unknown and may have varied across human populations. It may seem reasonable to assume that in Western societies, where there is typically no shortage of food, nutrition would not be a major health concern. Nevertheless, nutrition is blamed for many illnesses in contemporary Western societies, such as diabetes and obesity[4]; concurrently, the incidence of nutrition-related diseases in the developing world is rapidly increasing.[5]

Can an evolutionary perspective help us to understand and interpret the links between nutrition and health that form a major focus of research in the biomedical sciences? There is little doubt that contemporary humans eat differently compared to the diet we encountered during much of our evolutionary history[6]: a typical modern diet consists of much higher proportions of processed carbohydrates and saturated fats, and lower protein content. This has given rise to the idea that contemporary diets are insufficient to meet our nutritional needs shaped by natural selection.[1,6] In this scenario, nutrition-related health issues arise, at least in part, because our bodies are no longer adapted to our environments: in other words, they are mismatched. While this hypothesis has attracted considerable attention, its theoretical and empirical basis is not conclusive. There are many facets of contemporary nutrition that differ from what we have experienced during our evolutionary history, yet only a subset of these are linked to ill health. Some diets that are likely evolutionarily novel for humans, such as exclusive vegetarianism, in fact appear to increase health and lifespan compared to ancestral omnivorous conditions.[7] Furthermore, not everybody on, for example, a Western high-fat and high-sugar diet develops nutrition-related diseases. Caution is therefore required when alluding to evolutionary adaptation as a means to explain human disease, as it is not always possible to predict the complexity of responses by simply referring to the mismatch between adaptations to an evolutionarily ancestral environment and contemporary environments.

Our aim here is not to attempt a more ambitious evolutionary analysis for the causes of non-communicable diseases. Instead, we make use of the extensive body of work in evolutionary ecology to discuss the adaptive rationale for some of the mismatch theories: when they apply and when application is problematic, and how well the evidence supports these ideas in other animal systems. We focus on three aspects that are particularly relevant for those interested in linking nutrition and epigenetics: evolution of life histories where conditions early in life have long-term effects, maternal and transgenerational effects, and population differences due to genetic divergence. Needless to say, the evidence that epigenetic modifications (in the narrow molecular sense) are involved in responses observed in the wild is only tentative and rarely, if ever, demonstrated. Nevertheless, we make reference to epigenetic mechanisms whenever we feel this is theoretically warranted or plausible on empirical grounds. Throughout, we highlight how the evolutionary perspective is used to provide insights to an understanding of human health, and we present specific case studies of the role of dietary restriction in extending human lifespan and evolutionary insights on the current obesity crisis. An important issue to keep in mind throughout is that adaptive models of evolution focus on how features relate to biological fitness, the main components of which are survival and reproductive output. Evolutionary theory is rather silent on the question of whether individuals with high fitness are likely to be "healthy" or not, although in many circumstances the two should be positively correlated.

We use a broad definition of "nutrition" as any aspect of food that affects an animal.[8] Ecologists have typically considered animal nutrition in such broad terms, focusing on how animals maximize the rate of energy intake or using single-currency measures of specific nutrients for particular functions, such as the use of carotenoids in sexual displays.[9] A growing number of studies are extending this perspective to consider how individuals optimize intake of several macro- and micronutrients, using an approach termed the Geometric Framework.[8] This approach is informed by the field of applied animal nutrition, which studies how animals balance protein and carbohydrate intake, and ingest essential vitamins and minerals required for healthy development.[1] Such an approach can explain why it might be adaptive to over-eat carbohydrates, for example, in order to achieve a diet that contains the optimal amount of protein. This multivariate approach to understanding animal adaptation to some extent raises concerns about the explanatory value of traditional evolutionary models, and can lead to new insights regarding our ability to predict circumstances under which nutrition will cause poor health.

Long-term Effects of Early-life Conditions

While an animal is growing, it requires access to nutrients to develop tissues and organs, to attain the appropriate body size and survive until reproduction, to compete with other individuals for access to resources or mates, and to reproduce successfully. It is uncontroversial that nutritional quality or quantity in early life can affect individual fitness. Indeed, the evolutionary ecology literature is replete with studies demonstrating that early-life nutrition, from conception to maturation, can have long-term consequences for trait expression, reproductive performance, and survival in natural populations.[10] This may often be because of effects on body size, but another candidate mechanism by which nutrition affects later phenotype is through environmentally induced changes in gene expression, or epigenetics. Indeed, the link between nutrition, epigenetics and lifespan has driven research programs attempting to unveil "epigenetic diets" (such as folate provision or caloric restriction) that will increase lifespan.[11]

There are two broad perspectives on the adaptive function of these long-term effects (Figure 1). On the one hand, individuals who receive poor nutrition early in life may be constrained to develop on a trajectory of slow growth even if this means they have lower survival and reproductive performance than their well-nourished conspecifics who experience a "silver spoon" effect.[12,13] Even if these poorly nourished individuals then experience a change in circumstance and increase their growth to compensate

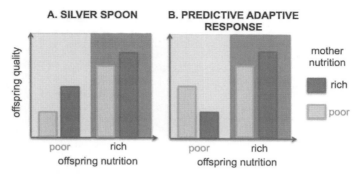

Figure 1. Simplified predictions of the two scenarios linking early-life or maternal nutrition to later health. A. "Silver spoon" scenario whereby individuals always do better when they have higher nutrition in early life; B. "Predictive adaptive response", or environmental matching, whereby individual quality is highest when the environment matches that of their early life or maternal cues. In both scenarios, the response to maternal nutrition is stronger when offspring themselves experience poor food conditions (pale boxes and background shading).

for their poor start, such catch-up growth can incur costs paid later in life.[14] On the other hand, individuals who start life under poor nutritional conditions may strategically grow slowly in order to develop a phenotype that is adapted to low food conditions (a form of plasticity often referred to as a "predictive adaptive response"[15,16]). In this scenario, individuals respond adaptively to cues about their anticipated later conditions. If these cues are mismatched to the realized conditions, individuals produce an inappropriate phenotype at a fitness cost. Both of these scenarios predict that early conditions can be associated with poor health later in life, and that the negative consequences depend on the environment (Figure 1). Below, we briefly outline the theoretical rationale for each of these scenarios, the strength of empirical support, and their relevance for understanding nutritional effects on human health.

Under the first scenario, the general assumption is that all individuals are striving to reach a particular phenotype, such as larger body size which often corresponds to higher reproductive success, and that individuals are responding within the constraints of their nutritional environment to maximize their survival and reproductive success (Figure 1A). Consider the fate of individuals who are born small because of limited resource availability during early development. How should such individuals adjust their growth, timing of maturation, and reproductive investment to maximize their fitness? Adjustments of growth may be possible but could be costly or constrained by a lack of resources.[17] If they do not adjust their growth, individuals that are small at birth may either mature at the same time as individuals that start out large, but at a smaller size, or delay maturation until they have reached the same size. Whether individuals should delay maturation will depend on costs of delayed development, for example increased exposure to predators during a vulnerable growth stage.[18] The impact of a poor start in life on growth and timing of maturation can thus have long-term consequences for reproductive success, as the size at which individuals mature commonly influences success at breeding (by determining, for example, the size and number of offspring that can be produced).[19]

Individuals who start out life under poor nutritional conditions often remain in a deprived environment throughout development. In many natural systems, however, the environment can fluctuate seasonally or unpredictably during development. An improvement in conditions presents an opportunity for individuals who were born small to increase their growth such that they are able to reach the same size without extending the developmental period. While this catch-up or compensatory growth can lead to individuals achieving similar reproductive success as their counterparts who started life in improved circumstances, there are a number of costs associated with rapid growth,[14] such as a build-up of cellular damage. In spite of these costs, there is widespread empirical support for catch-up growth across diverse taxa.[20] In humans, individuals who are born small for gestational age tend to grow rapidly when put on a high nutrition diet. While these individuals may have better health early in life, it is possible they will pay the cost for this growth spurt in terms of faster aging. Studies on laboratory rodents have shown, for example, that rapid catch-up growth following early nutritional deprivation results in higher biomarkers of aging and reduced lifespan.[21] This empirical evidence suggests that, in most cases, individuals are selected to utilize resources for growth even if this comes at a cost of lower survival prospects as a consequence of incurred physiological damage. Exactly what causes damage, and why individuals cannot protect themselves from it, is an area of active research and debate in functional ecology.[22] One attractive option is that metabolic processes by necessity generate oxidative stress. However, this inevitable "cost of living" leaves open the question of why there is such large variation among species in their ability to tolerate high metabolic rate.[23]

The second scenario that is invoked to explain the long-term effects of early nutritional conditions is that individuals are responding adaptively to cues in the environment of their later conditions.[24–26] In contrast to the previous framework, there are different optimal phenotypes to match particular nutritional conditions: rather than bigger always being better, it may be optimal to be small under low food conditions. A hallmark of adaptive plasticity is that individuals have highest fitness under conditions that most commonly follow — and are thus accurately predicted by — the environmental stimulus (Figure 1B).[27] As a consequence, individuals should suffer a fitness cost when cue and realized environment are mismatched. For example, when water fleas (Daphnia) sense the chemical cues of predators during development, they produce a protective helmet that is beneficial when there are predators but reduces mobility and is thus costly in the absence of predators.[28] Many other cases show that adaptive plasticity is common, but there are also many instances where plasticity would be adaptive but is not observed. For adaptive plasticity to evolve, the different environments need to occur with enough frequency such that there is selection for organisms to adjust their phenotype. Thus, plasticity can be absent even if not costly.

The mismatch between evolved adaptations and contemporary environments is used to explain the development of some non-communicable diseases associated with nutrition in humans, such as diabetes, obesity and cardiovascular disease.[29] Whether the mismatch perspective can provide insights on these phenomena requires them to be closely examined in light of the theoretical framework outlined above. Both scenarios can explain the long-term effects of early-life experiences of harsh environments even when later conditions improve: the incidence of later disease for individuals whose diet has switched between early and later life may be due to early-life cues being mismatched or due to costs of adjusting the growth trajectory. To demonstrate adaptive plasticity to local conditions, rather than life-history costs of growth, it is necessary to show that the environmental cues early in life are predictive of that experienced later, that there are different trajectories to specific phenotypes optimized in the different environments, and that the different environments occur often enough to select for adaptive plasticity.[27] These are demanding tests, particularly for humans. Consequently, the discussions regarding which, if any, of the adaptive responses that contribute to human disease are quite speculative.

Since tests of adaptive plasticity in humans are likely to remain challenging, we suggest that it is more urgent to assess if realistic assumptions about different adaptive responses are likely to be important enough to inform intervention or treatment policies. The predicted health differences between, for example, catch-up growth and predictive adaptive responses might be minor for the most obvious risk group: individuals born small and encountering a food environment rich in fat and sugar. A more relevant question in this context is how the developmental machinery translates the environment (e.g. dietary composition) into (mal)function. Using evolutionary theory to explicitly link ultimate and proximate causation could thus be a more promising avenue for evolutionary medicine than the current focus on adaptive "strategies", which provides very limited insight about mechanism.

Maternal and Transgenerational Effects of Nutrition

In the previous section, we presented two perspectives based on the role of the environment in providing resources or information for the developing organism that can generate long-term effects of early-life experiences. An important aspect of an individual's early environment is its mother, particularly for organisms such as mammals which spend the first part of their life developing inside the mother and relying exclusively on her for both resources and information about the outside world. Theoretical models on maternal effects revolve around these two different (but complementary) functions mothers play in their offspring's development — providing nutritional resources to offspring that can increase fitness directly and acting as a cue about the local conditions that the offspring can respond to adaptively.

Maternal nutritional effects may often have both functions; low nutrient provisioning limits offspring growth but also carries information about maternal nutritional status, which may affect the fitness benefits of, for example, natal dispersal.[30] As discussed above, such cues can also result in long-term effects when the environment is predictable across longer time scales. Other maternal effects have primarily resource or informational value. An example of the latter may be hormones, such as glucocorticoids and androgens. These can be transferred from mothers to offspring and appear to shape offspring development despite arguably having limited value as a source of energy or building blocks for somatic development, since the offspring in principle — if not in practice — could synthesize hormones themselves.[31]

There is a substantial body of theory on the role mothers play in shaping their offspring's development, which depends on the reliance of offspring on maternal resources, the predictability of environmental cues across generations, and the degree to which selection favors mothers to buffer their offspring or facilitate their rapid responses to environments.[31-34] A recent meta-analysis across plant and animal systems found weak overall support that offspring adaptively respond to their mother's environment.[35] It is more often the case that direct resources from the mother carry over to influence their offspring's development, not that maternal effects have been shaped to transmit information in a variable environment. One possible reason for this is that the maternal phenotype is often a poor cue to conditions experienced later in life or that such environments are too rare to allow evolution of adaptive plasticity. The number of studies that have quantified the information value of the maternal phenotype for offspring is very low, however, which makes it difficult to assess whether appropriate systems to test mismatch predictions have been identified.[35]

Whether mothers adaptively adjust their offspring's phenotype to be suited to a particular environment depends not only on the predictability of cues and whether mothers buffer their offspring, but also on the degree of evolutionary conflict between mothers and their young.[36] This conflict can arise because offspring — who express genes from their fathers as well as their mothers — may be selected to demand more resources from their mothers, at the expense of their siblings, than mothers are willing to provide. Parent-offspring conflict has been proposed to explain aspects of reproductive mode such as the structure of the placenta and how resources are transferred between mother and offspring at the placental barrier.[37] The conflict perspective has also been used to explain genomic imprinting, where the allelic copies from the mother and father have different epigenetic profiles, with one copy often being silenced. Paternal expression of the insulin-like growth factor gene *Igf2*, for example, enhances fetal growth whereas maternal expression of its associated receptor, *Igf2r*, has the reverse effect.[38] Although theoretically plausible, conflicting selection is not the only explanation; imprinting can also evolve to facilitate coadaptation when there is no evolutionary conflict between mother and father.[39] Indeed, the comparative data on imprinting suggest that conflict over resource utilization is unlikely to be the only explanation for imprinted loci.[38,40] Imprinted loci contribute to severe metabolic diseases in humans,[41] but we are not aware of studies that have contrasted conflict and non-conflict theories with respect to how they can be used to predict what, when, and how nutritional-related disease will be expressed in contemporary populations. (See Ref. 39 for a general discussion of the defining features of the imprinting theories.)

Environmentally induced transgenerational effects can occur through maternal nutrition influencing epigenetic marks on their offspring's DNA, such as modification of the patterns of DNA methylation. Indeed, the imprinting described above is a special case of such epigenetic effect. In a study in rural Gambia, for example, a mother's diet during pregnancy has been linked to the expression of a gene associated with immune function in her children.[42] Extensive studies of offspring born after the Dutch famine show that poor maternal nutrition can have effects on a number of genes associated with metabolic regulation, especially if offspring were conceived during the famine; effects later in development appear to be less pronounced.[43] Thus, maternal condition can influence gene expression profiles of their adult offspring. If these epigenetic marks are not erased from one generation to the next, they may influence subsequent generations. It has been shown that adding methyl donors, such as folates and vitamin B12, to the diet of pregnant mice affects the germline methylation status of the *agouti* gene, which determines coat patterning.[44,45] More recent experimental studies of both maternal and paternal exposure have showed that many more DNA regions can escape reprogramming.[46] Whether or not epigenetic inheritance is responsible for observed transgenerational effects of nutrition in wild animals is an important question for the future.

Is this incomplete resetting in the germline a simple glitch in the resetting mechanism, or are there circumstances under which it is adaptive? There has been very little theory on the conditions under which natural selection would favor incomplete erasure of epigenetic marks. A recent model showed that incomplete erasure can be adaptive when the environment is quite stable across generations and when maternal and offspring cues about the environment are unreliable.[47] Under these circumstances, passive inheritance of acquired epigenetic marks provides information about the environment over and above the information provided by direct environmental and maternal effects alone. It is currently unknown if species that live in such environments have a higher incidence of incomplete epigenetic resetting.

Mothers also transmit a diversity of bacteria to their young, many of which are crucial for the digestion of food.[48] For example, in obligate blood-feeding insects like tsetse flies, the transmission of bacterial symbionts is necessary for the uptake of essential B-vitamins not available through ingesting blood alone.[49] The evolutionary predictions on how mothers might adjust their offspring's microbiome depending on changing environmental conditions have yet to be formulated. The degree of coadaptation and conflict could be even more intense than that between mother and offspring alone, as a genome from a new type of organism is now in the picture. We expect that theoretical work on the evolution of the "holobiont"[50] will be helpful for predicting the role of microbiota for environmental and transgenerational effects on non-communicable diseases.

Genetic Divergence and Local Adaptation to Nutrition

Although developmental mechanisms are in place to ensure flexibility in response to changing nutritional needs, as explained above, a sustained environmental change, and limited dispersal between environments, can lead to local genetic adaptation.[51] Similarly, variation among populations in the rate of environmental change and the reliability of cues about nutrition can lead to genetic divergence in adaptive plasticity between or across generations (see above).

There is substantial evidence for local adaptation to diet in animals. For example, populations of Daphnia from lakes containing a diet high in toxic cyanobacteria have higher tolerance than populations that are not regularly exposed to the toxin[52] and transferring lizards to islands where the diet contains more plant material resulted in evolutionary divergence in bite force and gut morphology within a few decades.[53] In humans, a clear case of genetic responses to utilize new sources of nutrition can be seen in the local adaptation of populations to dairy farming. Phylogenetic analyses have shown that populations with a history of farming cattle have evolved tolerance of lactose beyond the age at which infants are weaned and there is indeed measurable selection on genes associated with digesting lactose.[54,55] At the same time, there has been co-evolution of genes associated with milk production in cattle.[54,55] Whereas the adaptive significance of these patterns are clear, broader evidence of adaptive genetic divergence in the ability to digest and convert different nutritional resources into growth and reproduction appear surprisingly rare given the large ecological differences between human groups, from culturally vegetarian populations to those which have experienced heavily meat- or fish-based diets for generations.

If adaptive genetic change is indeed limited, this may reflect how nutritional adaptation is partly driven by changes in gut microbiota, which limits selection on the human genome. Indeed, across mammals, the diversity of gut microbiota reflects variation in diet[56] and there are differences in the microbiome of children from European and African populations.[57] With the recent global migration patterns, which expose different ethnic and genetic groups to a range of environments, comparative tests of nutritional adaptation in humans that can exclude non-genetic inheritance are now possible. Experimental and correlative studies of wild animals have just begun and we anticipate that evidence for (co)adaptation via genetic and symbiont responses to ecological conditions will soon be available.

Evolutionary Insights on Current Public Health Issues: Living Longer and Dealing with Obesity

We have described three processes in which an evolutionary perspective can give insights on the link between nutrition and health in humans, depending on how individuals utilize resources during development, how mothers influence their offspring, and how sustained environmental challenges affect gene frequencies. We now present two special case studies where this evolutionary perspective has been used to generate insights into topics of public health concern: increasing lifespan and the obesity epidemic in the Western world.

Eat Less to Live Longer: An Evolutionary Perspective on the Link between Dietary Restriction and Lifespan

Across a range of organisms, from single-celled yeast to non-human primates, individuals who undergo mild but prolonged dietary restriction show a reduced incidence of age-related disease and live longer.[58] Initially it was thought that a restriction on calories was the mechanism underlying this phenomenon, but there is growing evidence that it is more likely to be driven by a shift in the balance of macronutrients.[59] In fruit flies (*Drosophila* species), for example, caloric restriction alone did not extend lifespan but individuals lived longer when experiencing a specific protein-to-carbohydrate ratio in their diet.[60]

From an evolutionary perspective, a health benefit of restricted diets might seem surprising, given that one might expect organisms to ingest as much food as possible to safeguard against harsh times. However, standard theory of resource allocation predicts that, when there is a reduction in the total amount of resources, individuals should invest less in costly reproduction or growth and more in somatic maintenance.[61,62] As a result, individuals experiencing a restricted diet may have longer lifespans because they have up-regulated their cellular repair mechanisms. This resource allocation hypothesis can explain sex differences in the link between diet and lifespan in species such as crickets, where males and females differ in how they optimize the trade-off between reproduction and maintenance.[63]

This simple life-history explanation for the link between dietary restriction and lifespan, although intuitively appealing, has not drawn the line under our understanding of the evolution of this phenomenon. First, it does not necessarily explain why the effects of dietary restriction are conserved across taxa with such diverse reproductive strategies and life histories. This might weaken the need for an adaptive explanation, as life extension in response to low food does not seem to be a mechanism that has evolved in response to species-specific environments. Second, most of the empirical evidence for lifespan extension after dietary restriction comes from model organisms which have been bred for many generations in the laboratory.[64] Indeed, whether dietary restriction has any effect under natural conditions is an open question.[65] Only with more nutritionally explicit theoretical models and a closer understanding of these mechanisms in the wild are we likely to resolve what insights evolutionary biologists can bring to this increasingly studied mechanism to achieve life extension.[1]

Environmental Mismatch and Obesity in Modern Humans

A major cause of concern to human health is the growing incidence of obesity across the world and the associated increase in age-related disease and morbidity. There are several evolutionary explanations for the obesity epidemic, unified by the theme that our current diet is mismatched from that of our evolutionary past.[66] Here, we describe two specific formulations of this hypothesis and the strength of support they have received.

In the 1960s, Neel proposed the "thrifty genotype" idea: that certain genes function to store excess fat so that we can survive through periods of famine, which results in obesity when these genes are expressed in the constantly high-nutrient conditions of modern times.[67] Two main problems with this perspective are that, first, it does not explain why the propensity to develop obesity is not general across populations.[68] Second, given that diets have progressively changed since the Paleolithic, and there is a great diversity of historic diets, it is not clear why there has not been subsequent selection against these so-called thrifty genes. Indeed, there are several studies showing that organisms can adapt to new diets experienced constantly across generations. Caterpillars, for example, fed high starch plants for multiple generations evolved resistance to the costly outcome of storing too much fat.[69]

A more recent type of mismatch explanation is based on how we achieve the optimal balance of macronutrients in our diet. Using this Geometric Framework perspective, Simpson and colleagues have argued that we are limited by the amount of protein in our diets, and in order to attain the necessary protein we over-ingest carbohydrates.[1] Indeed, comparative data from laboratory rodents and chickens shows that animals respond to a protein-poor diet by increasing their intake of carbohydrates, and contemporary human diets are certainly lacking in the relative protein content that would have been present in our diet in the past.[1] Theoretical models based on the Geometric Framework[70] will help to develop predictions on how such responses will depend on the strength of selection against obesity and constraints on adjusting the optimal protein intake.

The concept of evolutionary mismatch may also be approached from the reverse direction, i.e. by asking why it is that some phenotypes (and functions) are highly robust in the face of huge differences in nutrition.[71,72] Humans are a good example, because among healthy people, there is great variation in diet from those who consume animal products almost exclusively to vegan diets. This flexibility is often attributed to an evolutionary history of omnivory, i.e. that we have evolved to be diet generalists. What are the consequences of this for mismatch theories? Vegetable matter should consistently be suboptimal for members of a species with an evolutionary history of carnivory, since individuals should not have evolved the means to properly feed, digest, or process nutrients in plants. The consequences of a persistent vegetable diet for members of a species with a history of omnivory (such as humans) are more difficult to predict. The dietary environment is multivariate — composed of multiple variables, such as fats, sugars, and carbohydrates — and variable in amount and composition on different scales, both within and between generations. Moreover, there is variation in nutritional demand both within and among individuals, such that the fitness consequences of nutrition are context-dependent, for example, depending on reproductive status. These complexities make it difficult to predict the dietary changes individuals will be able to tolerate, with and without adjusting their physiology, as well as when diet will compromise fitness and health. Incorporating concepts from evolutionary biology on the robustness of biological systems[71–73] may bring some clarity to these complex predictions in evolutionary medicine.

Summary and Conclusions

In this chapter, we have outlined three main processes in which an evolutionary perspective linking nutrition and health may generate insights useful for the biomedical sciences. We have described how theories linking nutrition to development can conceptualize these effects either as providing direct resources or cues to the organism; they can be passed on across generations; and consistent exposure to varied nutrition between populations can lead to genetic divergence. There is a growing body of theoretical models describing the systems in which these different scenarios take place and considerable interest in whether they apply to humans. Whether an evolutionary perspective can lead to insights on the link between diet, development and health in humans depends on the extent to which the assumptions of the models fit the reality of human life history and environmental regimens. We suggest that a fruitful approach is to not only focus on the evolution of life-history strategies in the traditional adaptationist program but on the evolution of mechanisms. This may facilitate predictions for what features of contemporary environments, among all those that are evolutionary novel, will make us sick.

Acknowledgements

The authors received financial support from the European Union's Seventh Framework Programme (FP7/2007–2011) under grant agreement no. 259679 (to T.U.). T.U. and S.E. are supported by the Royal Society of London, and T.U. by the Knut and Alice Wallenberg Foundation. We thank Bram Kuijper and Becky Kilner for useful discussion.

References

1. SJ Simpson, D Raubenheimer. (2012) *The Nature of Nutrition: A Unifying Framework from Animal Adaptation to Human Obesity*. Princeton: Princeton University Press.
2. D Pfennig. (1990) The adaptive significance of an environmentally-cued developmental switch in an anuran tadpole. *Oecologia* **85**:101–107.

3. D Schluter. (2001) Ecology and the origin of species. *Trends Ecol Evol* **16**:372–380.
4. L Cordain, SB Eaton, A Sebastian, N Mann, S Lindeberg *et al.* (2005) Origins and evolution of the western diet: health implications for the 21st century. *Am J Clin Nutr* **81**:341–354.
5. BM Popkin, LS Adair, SW Ng. (2012) Global nutrition transition and the pandemic of obesity in developing countries. *Nutr Rev* **70**:3–21.
6. SB Eaton, SB Eaton III, MJ Konner, M Shostak. (1996) An evolutionary perspective enhances understanding of human nutritional requirements. *J Nutr* **126**:1732–1740.
7. CT McEvoy, N Temple, JV Woodside. (2012) Vegetarian diets, low-meat diets and health: a review. *Public Health Nutr* **15**:1–8.
8. D Raubenheimer, SJ Simpson, D Mayntz. (2009) Nutrition, ecology and nutritional ecology: Toward an integrated framework. *Funct Ecol* **23**:4–16.
9. TD Price. (2006) Phenotypic plasticity, sexual selection and the evolution of colour patterns. *J Exp Biol* **209**:2368–2376.
10. J Lindström. (1999) Early development and fitness in birds and mammals. *Trends Ecol Evol* **14**:343–348.
11. MG Bacalini, S Friso, F Olivieri, C Pirazzini, C Giuliani *et al.* (2014) Present and future of anti-ageing epigenetic diets. *Mech Ageing Dev* **136**:101–115.
12. A Grafen. (1988) On the uses of data on lifetime reproductive success, in *Reproductive Success: Studies of Individual Variation in Contrasting Breeding Systems,* TH Clutton-Brock, Editor. Chicago: University of Chicago Press, 454–471.
13. P Monaghan. (2008) Early growth conditions, phenotypic development and environmental change. *Phil Trans R Soc B* **363**:1635–1645.
14. M Mangel, SB Munch. (2005) A life-history perspective on short- and long-term consequences of compensatory growth. *Am Nat* **166**:E155–E176.
15. P Bateson, D Barker, T Clutton-Brock, D Deb, B D'Udine *et al.* (2004) Developmental plasticity and human health. *Nature* **430**:419–421.
16. PD Gluckman, MA Hanson, HG Spencer, P Bateson. (2005) Environmental influences during development and their later consequences for health and disease: implications for the interpretation of empirical studies. *Proc R Soc B* **272**:671–677.

17. CM Dmitriew. (2011) The evolution of growth trajectories: what limits growth rate? *Biol Rev* **86**:97–116.
18. V Remes, TE Martin. (2002) Environmental influences on the evolution of growth and developmental rates in passerines. *Evolution* **56**: 2505–2518.
19. DA Roff. (1992) *The Evolution of Life Histories: Theory and Analysis.* New York: Routledge, Chapman and Hall.
20. KL Hector, S Nakagawa. (2012) Quantitative analysis of compensatory and catch-up growth in diverse taxa. *J Anim Ecol* **81**:583–593.
21. JL Tarry-Adkins, SE Ozanne. (2014) The impact of early nutrition on the ageing trajectory. *Proc Nutr Soc* **73**:289–301.
22. C Selman, JD Blount, DH Nussey, JR Speakman. (2012) Oxidative damage, ageing, and life-history evolution: where now? *Trends Ecol Evol* **27**:570–577.
23. C Isaksson, BC Sheldon, T Uller. (2011) The challenges of integrating oxidative stress into life-history biology. *Bioscience* **61**:194–202.
24. TW Fawcett, WE Frankenhuis. (2015) Adaptive explanations for sensitive windows in development. *Front Zool* **12**(Suppl 1):S3.
25. WE Frankenhuis, K Panchanathan. (2011) Balancing sampling and specialization: an adaptationist model of incremental development. *Proc R Soc B* **278**:3558–3565.
26. S English, TW Fawcett, AD Higginson, PC Trimmer, T Uller. (2016) Adaptive use of information during growth can explain long-term effects of early life experiences. *Am Nat* **187**:620–632.
27. P Doughty, DN Reznick. (2004) Patterns and analysis of adaptive phenotypic plasticity in animals, in *Phenotypic Plasticity: Functional and Conceptual Approaches,* TJ DeWitt, SM Scheiner, Editors. New York: Oxford University Press, 126–150.
28. AA Agrawal. (2001) Phenotypic plasticity in the interactions and evolution of species. *Science* **294**:321–326.
29. PD Gluckman, MA Hanson, P Bateson, AS Beedle, CM Law et al. (2009) Towards a new developmental synthesis: adaptive developmental plasticity and human disease. *Lancet* **373**:1654–1657.
30. M Massot, J Clobert. (1995) Influence of maternal food availability on offspring dispersal. *Behav Ecol Sociobiol* **37**:413–418.
31. MJ Sheriff, OP Love. (2013) Determining the adaptive potential of maternal stress. *Ecol Lett* **16**:271–280.

32. S English, I Pen, N Shea, T Uller. (2015) The information value of non-genetic inheritance in plants and animals. *PLoS One* **10**:e0116996.
33. B Kuijper, RB Hoyle. (2015) When to rely on maternal effects and when on phenotypic plasticity? *Evolution* **69**:950–968.
34. O Leimar, JM McNamara. (2015) The evolution of transgenerational integration of information in heterogeneous environments. *Am Nat* **185**:E55–E69.
35. T Uller, S Nakagawa, S English. (2013) Weak evidence for anticipatory parental effects in plants and animals. *J Evol Biol* **26**:2161–2170.
36. T Uller, I Pen. (2011) A theoretical model of the evolution of maternal effects under parent-offspring conflict. *Evolution* **65**:2075–2084.
37. B Crespi, C Semeniuk. (2004) Parent-offspring conflict in the evolution of vertebrate reproductive mode. *Am Nat* **163**:635–653.
38. D Haig. (2004) Genomic imprinting and kinship: how good is the evidence? *Annu Rev Genet* **38**:553–585.
39. MM Patten, L Ross, JP Curley, DC Queller, R Bonduriansky *et al.* (2014) The evolution of genomic imprinting: theories, predictions and empirical tests. *Heredity* **113**:119–128.
40. C Köhler, I Weinhofer-Molisch. (2010) Mechanisms and evolution of genomic imprinting in plants. *Heredity* **105**:57–63.
41. KD Robertson. (2005) DNA methylation and human disease. *Nat Rev Genet* **6**:597–610.
42. MJ Silver, NJ Kessler, BJ Hennig, P Dominguez-Salas, E Laritsky *et al.* (2015) Independent genomewide screens identify the tumor suppressor VTRNA2-1 as a human epiallele responsive to periconceptional environment. *Genome Biol* **16**:118.
43. EW Tobi, JJ Goeman, R Monajemi, H Gu, H Putter *et al.* (2014) DNA methylation signatures link prenatal famine exposure to growth and metabolism. *Nat Commun* **5**:5592.
44. JE Cropley, CM Suter, KB Beckman, DIK Martin. (2006) Germ-line epigenetic modification of the murine A vy allele by nutritional supplementation. *Proc Natl Acad Sci USA* **103**:17308–17312.
45. JE Cropley, THY Dang, DIK Martin, CM Suter. (2012) The penetrance of an epigenetic trait in mice is progressively yet reversibly increased by selection and environment. *Proc R Soc B* **279**:2347–2353.

46. S Seisenberger, JR Peat, TA Hore, F Santos, W Dean et al. (2013) Reprogramming DNA methylation in the mammalian life cycle: building and breaking epigenetic barriers. *Phil Trans R Soc B* **368**:20110330.
47. T Uller, S English, I Pen. (2015) When is incomplete epigenetic resetting in germ cells favoured by natural selection? *Proc R Soc B* **282**:20150682.
48. F Ponton, K Wilson, SC Cotter, D Raubenheimer, SJ Simpson. (2011) Nutritional immunology: a multi-dimensional approach. *PLoS Pathog* **7**:e1002223.
49. BL Weiss, J Wang, S Aksoy. (2011) Tsetse immune system maturation requires the presence of obligate symbionts in larvae. *PLoS Biol* **9**:e1000619.
50. SF Gilbert, J Sapp, AI Tauber. (2012) A symbiotic view of life: we have never been individuals. *Q Rev Biol* **87**:325–341.
51. TJ Kawecki, D Ebert. (2004) Conceptual issues in local adaptation. *Ecol Lett* **7**:1225–1241.
52. O Sarnelle, AE Wilson. (2005) Local adaptation of *Daphnia pulicaria* to toxic cyanobacteria. *Limnol Oceanogr* **50**:1565–1570.
53. A Herrel, K Huyghe, B Vanhooydonck, T Backeljau, K Breugelmans et al. (2008) Rapid large-scale evolutionary divergence in morphology and performance associated with exploitation of a different dietary resource. *Proc Natl Acad Sci USA* **105**:4792–4795.
54. A Beja-Pereira, G Luikart, PR England, DG Bradley, OC Jann et al. (2003) Gene-culture coevolution between cattle milk protein genes and human lactase genes. *Nat Genet* **35**:311–313.
55. SA Tishkoff, FA Reed, A Ranciaro, BF Voight, CC Babbitt et al. (2007) Convergent adaptation of human lactase persistence in Africa and Europe. *Nat Genet* **39**:31–40.
56. RE Ley, M Hamady, C Lozupone, PJ Turnbaugh, RR Ramey et al. (2008) Evolution of mammals and their gut microbes. *Science* **320**:1647–1651.
57. C De Filippo, D Cavalieri, M Di Paola, M Ramazzotti, JB Poullet et al. (2010) Impact of diet in shaping gut microbiota revealed by a comparative study in children from Europe and rural Africa. *Proc Natl Acad Sci USA* **107**:14691–14696.

58. L Fontana, L Partridge. (2015) Promoting health and longevity through diet: from model organisms to humans. *Cell* **161**:106–118.
59. MDW Piper, L Partridge, D Raubenheimer, SJ Simpson. (2011) Dietary restriction and aging: a unifying perspective. *Cell Metab* **14**:154–160.
60. KP Lee, SJ Simpson, FJ Clissold, R Brooks, JWO Ballard *et al.* (2008) Lifespan and reproduction in Drosophila: new insights from nutritional geometry. *Proc Natl Acad Sci USA* **105**:2498–2503.
61. DP Shanley, TB Kirkwood. (2000) Calorie restriction and aging: a life-history analysis. *Evolution* **54**:740–750.
62. C Hou, K Bolt, A Bergman. (2011) A general life history theory for effects of caloric restriction on health maintenance. *BMC Syst Biol* **5**:78.
63. AA Maklakov, SJ Simpson, F Zajitschek, MD Hall, J Dessmann *et al.* (2008) Sex-specific fitness effects of nutrient intake on reproduction and lifespan. *Curr Biol* **18**:1062–1066.
64. S Nakagawa, M Lagisz, KL Hector, HG Spencer. (2012) Comparative and meta-analytic insights into life extension via dietary restriction. *Aging Cell* **11**:401–409.
65. MI Adler, R Bonduriansky. (2014) Why do the well-fed appear to die young? A new evolutionary hypothesis for the effect of dietary restriction on lifespan. *BioEssays* **36**:439–450.
66. KM Godfrey, KA Lillycrop, GC Burdge, PD Gluckman, MA Hanson. (2007) Epigenetic mechanisms and the mismatch concept of the developmental origins of health and disease. *Pediatr Res* **61**:5R–10R.
67. JV Neel. (1962) Diabetes mellitus: a "thrifty" genotype rendered detrimental by "progress"? *Bull World Health Organ* **77**:353–362.
68. BL Turner, AL Thompson. (2013) Beyond the Paleolithic prescription: incorporating diversity and flexibility in the study of human diet evolution. *Nutr Rev* **71**:501–510.
69. J Warbrick-Smith, ST Behmer, KP Lee, D Raubenheimer, SJ Simpson. (2006) Evolving resistance to obesity in an insect. *Proc Natl Acad Sci USA* **103**:14045–14049.
70. AM Senior, MA Charleston, M Lihoreau, J Buhl, D Raubenheimer *et al.* (2015) Evolving nutritional strategies in the presence of competition: a geometric agent-based model. *PLoS Comput Biol* **11**:e1004111. doi:10.1371/journal.pcbi.1004111.

71. P Bateson, P Gluckman. (2011) *Plasticity, Robustness, Development and Evolution.* Cambridge: Cambridge University Press.
72. M Félix, A Wagner. (2008) Robustness and evolution: concepts, insights and challenges from a developmental model system. *Heredity* **100**:132–140.
73. TF Hansen. (2011) Epigenetics: adaptation or contingency? In *Epigenetics: Linking Genotype and Phenotype in Development and Evolution,* B Hallgrimsson B, BK Hall, Editors. Berkeley: University of California Press, 357–376.

The Body Politic: Epigenetics and Society

CHAPTER 10

Noela Davis

School of Social Sciences,
The University of New South Wales,
Sydney, NSW 2052, Australia

Introduction

In examining the implications that epigenetic mechanisms and modifications have for society, there are two main aspects that come to the fore in terms of relevance for how we, as a society, respond to the epigenetic changes wrought in bodies. These are: what impact do society and social actions have on gene expression, which is part of the question of bodily plasticity; and over what time period are these effects felt, that is, the issue of the intergenerational transfer of epigenetic changes. These matters have implications for how we conceptualize the individual, and their relation to society and to disease causality. This in turn has implications for the sort of policy interventions and lifestyle recommendations it is appropriate or possible to make.

This review will not be giving in-depth examinations of specific fields — such as health policy or nutritional recommendations, or of the molecular detail of epigenetic processes — but will take an integrative and interdisciplinary approach to look at what our growing knowledge of epigenetic operations can say about how we go about our daily lives, and about the implications — policy, practice and ethics — of this for those who would attempt to regulate and fund health and other programs of social intervention.

I will not be making prescriptive suggestions as to how society can benefit from epigenetics, as there can be no one correct answer to the issues raised. Instead, these implications point to conversations in which a society needs to engage. The social ethos needs to be brought to bear on the results of epigenetic research to enable policy makers, political, social and medical agencies, and the general population to reach agreement as to what is an acceptable direction for *their* particular society.

Epigenetics

Epigenetics has sparked much interest as it is a discipline which offers the promise of an in-depth understanding of what it is to be human — much as the Human Genome Project did before it became apparent that its results raised as many questions as they answered. However, in its wake we have a greater appreciation of the complexity of life, and epigenetics appears to be helping unravel some of the mystery as it gains popularity as a narrative of our times. But even now it is evident that epigenetic research presents us with many ambiguities and puzzles of its own.

With epigenetics gaining in popularity in the public imagination, several recent studies[1-4] have examined how the social sciences and the popular media use epigenetic research to comment on our social selves or to offer popular science explanations of medical and social issues facing us today. A prominent finding of these reviews is that there seems to be confusion as to what, exactly, epigenetics is, even amongst scientists. The writers note that, while most research articles surveyed would give a variation on the standard definition, that is, that epigenetics relates to heritable changes in gene expression that cannot be explained by changes in DNA sequence,[5] often the results reported in the research would not neatly fit a simple account of what epigenetics "is". But many researchers do not comment on the ambiguities that manifest in their findings. The writers contend that even when scientists note that epigenetics contests the notion that we can separate and isolate, or molecularize, aspects of existence, they will, at the same time, engage in just such an isolation. Admittedly, as a heuristic device under scientific methodology, it is necessary to hold most aspects constant to study them. However, the discussion and conclusions of research could benefit by acknowledging this issue and resituating research, but many do not recontextualize their findings — and as I will argue, epigenetics revolves around context.

Stelmach and Nerlich[1] elaborate on the importance of recontextualization to remind us that the way we use terms and characterize the processes involved by use of metaphor or comparison shapes the way we think and act, and how we construct scientific fields. This, in turn, will have implications for how we put scientific knowledge to social use and construct the political and social narratives to justify any proposals.

While I cannot offer any neat solution to the definitional puzzle, perhaps the confusions noted arise from the very nature of epigenetics, for it is an ambiguous process. Epigenetics blurs boundaries, such as those between nature and nurture, nature and culture, body and society, organism and environment. But this destabilization of boundaries is not a reductionism or a claim that the two sides of these oppositions are the same, because, as Meloni[6] insists, difference and distinction certainly remain. We could even, as I will elaborate later, profitably consider epigenetics *as* differentiation and boundary making. We could envision it as a series of convoluted operations that have an impetus to take the pluripotential of the embryonic cell and divide and differentiate this into all the various tissues, organs and structures of the body at the same time as it also both defines a body in its environment and embodies this very environment.

But more concretely, epigenetics may be broadly described as the molecular processes of "phenotypic induction" through the interactions of individual bodies with their environmental milieu.[7] Even though it is common to read of these processes referred to as "programming", Burdge and Lillycrop[7] caution against using this term because of its deterministic connotations. For one thing that epigenetics shows us is the complexity and non-predictability of the outcome in any given instance. Epigenetic processes give an individual his or her specific propensities, suggest probable life-course outcomes, but cannot be used to predict these with certainty because epigenetic interactions with an organism's environment are contingent and always occur in a specific context. As Rose[8] notes, because of epigenetics' probabilistic nature there is no single future written in the present. Outcomes depend on what is experienced, when and how, and on how the phenotypic profile — which is *already* a product of an individual's responsiveness to its surroundings — responds to and processes environmental stimuli. These are procedures and reactions that can only be charted retrospectively due to epigenetic responses being built on previous environmental responses and proclivities.[9] In recognition of the non-determinacy of these operations, Burdge and Lillycrop sum up this body-environment conversation as the induction of a "differential risk of disease", but add the important rider that the majority of reactions in this individual-environmental porosity may be neutral.[7]

To enlarge on this, we should keep in mind that a differential risk of disease includes little probability of developing a disease, so epigenetics is as much about health as it is about disease. This is a point to keep in mind as most research will take disease processes as the object. Rose, drawing on Canguilhem, offers a comment which may illuminate this feature of research. Health is portrayed as "life lived in the silence of the organs", as opposed to disease, which is that which irritates and thus demands attention.[8] Health can take many forms and as it does not impinge upon or impede us, we may take it for granted. Disease and dysfunction, on the other hand, make themselves known in specific ways that can make them easier to define. But the same epigenetic processes — those of body-environment contextual conversation — underlie both, and in this way disease can give insights into the everyday. We should conceptualize epigenetics, not as a disease process, but a process of life where all outcomes on the disease-good health spectrum are the results of contingent epigenetic processes of individual specificity in a continuing conversational interplay with the particular milieu.

Nor are epigenetic adaptations and adjustments good or bad in themselves: it is only in the context of the particular life circumstances that we can assign the labels of detrimental, neutral or beneficial. For instance, while stress, and in particular a heightened stress responsivity to stimuli, is detrimental to health, in certain circumstances it is a useful way to approach one's surroundings. If you, for example, live on the streets, or are a member of a despised minority, being hypervigilant can help you avert danger or threat, and in so doing has its benefits, in the short term at least.[10–12]

Responsiveness to Surroundings

Epigenetic bodies have a complex relationship with both their physical and social surroundings and their history. They exhibit, at the same time, both plasticity and stability, an immediate responsiveness to milieu together with an intergenerational transfer of ancestral environmental receptiveness. It is now generally accepted that the epigenetic responses of a body to its environment have the potential to be passed on to, and manifest biologically in, subsequent generations, and this time-scale is an important factor to consider when looking at the implications of epigenetics for society. Research also suggests that it is not just physical exposures, for example, to toxic chemicals or to famine, that can produce intergenerational health effects, but that social events, such as stigma, discrimination and socio-economic disadvantage,[12–14] can bequeath a propensity to illness in the same ways.

At the same time, bodies are continually open to environmental influences and are involved in an ongoing conversational sampling of their milieu. Several recent human studies have demonstrated this ongoing openness by tracking real-time epigenetic changes in subjects as they are exposed to various conditions.[15–17] But even though the body can respond to environmental exposures instantaneously, this does not necessarily mean that this exposure straightforwardly leads to any immediate consequences. Stabilization of genetic expression in response to exposures can take different time periods depending on a host of factors such as the time of life at which the contact occurred, which epigenetic mechanism was implicated, which tissue was involved, as well as that body's established pattern of responsiveness — that is, its induced epigenetic profile.[18]

In an attempt to understand what might be happening in respect to a body's interactions with its environment and it history, Kuzawa and Thayer[19] have developed the hypothesis of "phenotypic inertia".[a] They contend that through our continuing body-environment communications, biology calculates a running average of the conditions it meets, as it were sampling them and adding to the body's historical epigenetic "database". They explain that the body accumulates and remembers its exposures and the responses it makes, and calibrates the epigenome to buffer unstable environmental features. That is, the body calculates long-term trends and has the ability to recognize, and thus ignore, transient spikes. To become embodied, an environmental condition must constitute a sustained and consistent encounter to allow it to become incorporated into the body's biological trend line, and to thus potentially be passed on to the next generation. In this way we can see that the body, as Kuzawa and Thayer say, works on intergenerational time-scales. This, in turn, implies that social, political or medical interventions that we, as a society, decide to make to promote health also need to be sustained and consistent to work in this biological time frame.

These epigenetic bodies are not merely passive meeting points for the internal and external environment[20] but are active agents in these conversational transactions. There is a mutual solicitation between body and environment[6] where responses are always specific to the particular qualities of both body and context. Bodies are always open to environmental influences and the potential changes that these can bring about, yet also maintain their identities, their particular propensities and ways of responding to the milieu in the

---

[a]The suggestion of phenotypic inertia should be understood only on an intergenerational time-scale and not situated in a geological time frame. That is, any hypothesized adaptive lag between bodies and environments is not that propounded by, for instance, the currently popular paleo diet movement. Research suggests there will not be a prehistoric body that has persisted unaltered throughout our history of cultural, agricultural and nutritional change.[19,26]

face of these events. There is a temporal and spatial complexity within the epigenetic body that contests any simple or linear notion of causality or of future-oriented determinism. This, as we can appreciate, presents a complex implication of individual and environment that cautions against attempts to predict or determine outcomes with any certainty in advance of the exposure and response in question.

Nutrition and Metabolism

Food is not a neutral or passive substance, nor does it necessarily affect individuals in the same way, or even have the same effects on different tissues in a single body. It is not simply fuel for the body but is a bio-active constructor of bodies and behaviors.[21] Food already embraces many environmental and social exposures attendant to its production, for instance, the degree of industrialization, farming practices, pesticide or antibiotic use, abundance or famine, whether it is presented to us processed or unprocessed, as well as social norms surrounding our customary eating habits, and these already embedded practices have the potential to impact upon bodies. In recognition that food itself already manifests the environment of its production Landecker conceptualizes eating as the simultaneous ingestion of the material and the social.[21] With food already being a material-social entanglement it is no surprise that research into the epigenetics of nutrition has revealed that food's effect on the body is extremely complex. The way food is processed by a body, and the outcomes of this processing, depend on the epigenetic profile that has been induced by the individual's specific history of environmental exposures acting with the practices implicated within the food itself, in its particular time and place.

Nutrition and metabolic processes are not simply the ingestion and digestion of food and the transport of metabolites through the body but are complicated environmental interactions. It matters whether a nutrient is, for instance, consumed in the context of a whole food or as a supplement; it makes a difference how food is processed — which particular additives, fats, sweeteners and preservatives it includes; it matters how much of it is consumed and by whom. It can also matter whether the consumer is male or female[22]; or at what time of life the exposure is encountered.[23] Quantities, too, are important, because even with known beneficial supplements, the therapeutic dose can be critical — too much or too little may be detrimental rather than curative.[24]

It also matters about the specific characteristics, propensities and responses of the individual who consumes, as the epigenetic effects of nutrition are always contextual. Different individuals, each with their own particular epigenetic inheritance, may respond to the same nutrient element, administered in the same concentration, differently. Each metabolism differs, as it is not simply a factory to digest and convert food but is, according to Landecker, a "regulatory interface shaped by the environment".[21] Rönn and Ling,[16] for instance, contend that exercise, among its effects on the body, molds metabolism by increasing the influence of muscle tissue, relative to adipose tissue, on the digestive system. Both these tissue types play their own roles in the metabolism, specific transport routes, and uses of various metabolites in the body. This adds further complexity to the question of the effects of nutrition, as the way any particular body metabolizes its nutrient intake will replay back into the epigenetic status of that body.

Food, as two recent studies[25,26] suggest, can also both shape metabolism and modify behavior. A study by Luo et al.[25] compared the effects of glucose versus fructose on metabolism and behavior, and while it does examine nutritional elements in isolation rather than in the context of complete foods, it provokes interesting questions for further study. The experiment was testing the ability to delay gratification in response to the ingestion of these sugars. Glucose is metabolized through insulin-secreting, satiety-signaling systems while fructose takes a different pathway through the body and does not promote feelings of satiety. After consuming these sugars, subjects were offered the choice between food now or money later. Those in the glucose group were much more likely to be able to wait for the monetary reward whereas those in the fructose group were significantly less able to delay gratification and tended to select the immediate food option.

What we can draw from this preliminary investigation is that food may influence other food choices: the subjects given fructose are eating when they, by other measures, don't need to. Such a practice opens the possibility of a greater risk of metabolic syndromes and also suggests economic consequences, as these subjects valued food and money differently to those consuming glucose.

In her review of research into dieting programs, Sanabria[26] posits a similar relationship between food, metabolism and the environment. Her investigations support the commonly held view that dieting is a largely unsuccessful exercise for many people, with any weight lost seeming to inevitably return. However, she doesn't attribute this to a lack of willpower or motivation in dieters, but to their bodily composition, to how many, and what type of, fat cells they have. Failure of diets is not because of personal inadequacies but because, according to the narrative of the studies, adipocytes have their own agency: fat cells are described as hungry, and they have apparently developed an affinity with our current culture of readily available, palatable junk food. In concert with this is an account of the agency of food, where food also directs our further choices. In the works that Sanabria cites, both food and fat appear to be resistant to any "mind over matter" calls for dietary restriction in the individual.

This leads Sanabria to pose a question regarding the narrative of mismatch whereby our bodies supposedly cannot adjust fast enough to the change in food type and availability, resulting in the current epidemic of metabolic syndromes. What if, she ponders, this is not an accidental occurrence, or a mismatch at all? What if we — or should that be our fat cells? — are complicit in this and have helped create the very conditions and foods which are allegedly making us ill?

However, at the same as this convoluted intricacy is being demonstrated in the relationship between individuals, foods and environments, there is also a tendency evident in some research to molecularize, to examine events at the molecular level, rather than to contextualize. Critiques of this habit[6,8,21,26,27] contend that this leads to a reduction of complexity surrounding food, a lack of attention to the specificity of body and environment, with the assumption that all forms of a nutritional element are equivalent, and thus will have the same effect in the body across different individuals. Food, and epigenetic processes and outcomes, become reduced to a single meaning,[26] and what is missed in this sort of analysis are the potential behavioral and social effects. Rather than examine the individual, their whole diet, and their life context, some subscribe to a "logic of substitution".[21] This is a logic that underlies the practices of dietary supplementation and the synthezising of nutrition — for instance, the notion that the nutrients from an energy bar are the same as the whole food from which the various amino acids, vitamins and carbohydrates came, and are processed by the body in the same way.[21] Or, indeed, the notion that one no longer needs to eat conventional food at all and can instead substitute completely synthesized products such as Soylent for some or all meals.[28]

While we don't have the evidence to say what the outcomes of long-term supplementation or substitution are, there is evidence that bodies know what they are consuming and isolated nutritional elements are not necessarily metabolized in the same way as the nutritional complexity of whole foods. This was brought into focus in some animal studies where counterintuitive results confounded expectations. Contrary to the hypothesis, a synthetic diet supplemented with methyl donors, and a synthetic diet stripped of methyl donors *both* lowered methylation.[21] This suggests that in assessing the effects of diet and nutrition, it is not simply the nutrient elements which have an effect but, rather, their complex interactions with all the components of a food and with the metabolism processing them that need to be taken into account.

These studies support Landecker's[21] contention that food is also environmental exposure, which in this case includes the social surroundings. They also raise the question of where the decision-making center is: who or what is deciding to forego money for food? Is it the person or the fat cell or, indeed, the food, directing the consumption of junk food? Where would we locate the source or agency of the decision? If the choice is influenced by the food consumed and the attendant metabolic processes, can we accurately claim that the person is solely or fully responsible for this choice?

Epigenetic Individuals

The labyrinthine interconnections and co-constitutiveness of both physical and social environments and biology, and between one generation and another, confound any attempts to maintain strict boundaries, such as those between self and other, the individual and their surroundings, or more generally nature and culture or nurture. In recognition of this entanglement Rose[27] describes epigenetics as the inseparability of vitality and milieu. He expands on this by arguing that responsivity, agency and responsibility are not properties of any single individual but are a distributed, implicated and meshed interactivity. Here we find not a one-way traffic but multiple interdependent webs of organic and social practices. Epigenetics can be visualized as one of life's processes that, as Hinton[29] contends, demonstrates that matter cannot be absolutely or properly separated, spatially or temporally, from its surroundings. In other words, as Chiew maintains, we are not dealing with self-contained entities but with the phenomenological and ontological implication of entities within other entities, in this case, bodies and environments as constitutive of each other.[30] Rather than a dualist conception of a passive body worked on by culture, epigenetic bodies raise the question of whether there can possibly be a natural biological life that is not *already* an openness to the social, and already cultural and political expression.[31]

Epigenetics thus suggests that borders are always porous and mutable. Jamieson,[31] to extend this analysis, contends that what we witness in processes such as epigenetics is not a blurring of boundaries or an infection of the biological by the social, but a demonstration that boundaries are *already* blurred. To add to this, the intergenerational responsiveness of individuals confuses dualist and neoliberal notions of self-contained and stable identity, of being comfortably able to locate a distinct, spatially and temporally delimited individual who is responsible and whose self-authored actions can be brought into play to resolve their own, or society's, problems. As Pitts-Taylor[32] elaborates, our prevailing view of the individual is tied up with a neoliberal, market-based ideology where we consider that the individual is in control of their actions, and good, moral, citizenship is linked with each taking personal responsibility for their own actions and their own health as they strive to not be a burden on the state and a drain on its resources. As a corollary, if individuals fail to look after themselves it becomes personal as it is assumed that they lack the proper willpower or have moral failings, and are thus in some way not exhibiting the characteristics of good citizenship.

But is this neoliberal notion what epigenetics suggests about the individual and their relationship to others and to society? Does epigenetic research into nutrition and health support the idea that individuals are the separately identifiable beings of dualist or neoliberal reasoning that can be fully in control of their own health and their own lives? If both your bodily and psycho-social propensities are induced by your own and your ancestors' history of environmental exposures, and by bodily responses to these, it does not seem that this prevailing model of the individual is appropriate to a discussion of sociality, responsibility and health. If, as I have discussed, there is evidence that food is not simply fuel for the body but is a shaper of bodies as it constitutes metabolisms, behavior and socioeconomic choices in particular ways, is it then possible to hold a person solely responsible for control of their life and their health?

Epigenetics suggests that the current incidence of metabolic syndromes and nutritionally induced ill-health could possibly have been induced up to several generations ago by the intricate web of bodily-environmental-historical cross-talk. And at the same time this enmeshing could have already induced various metabolic risks in generations to come. As well as these considerations there is also the suggested agency and complicity of foods and bodily tissues to take into account. These factors call into question that we can locate the individuals who are responsible for any disease, as there is a temporal and spatial involution that cannot be unraveled to allow us to pin-point an origin of disease or a negligent individual to blame.

Biopolitics as the Sphere of Intervention

How, then, can we as a society act to ameliorate nutritionally induced epigenetic changes that have been shown to increase the risk of ill-health if epigenetic effects are the implicated mesh of environmentally and temporally contingent interactions suggested by research? If there is no certain predictability or determinacy in epigenetics that can guarantee that the risk suggested by an individual's epigenetic profile will ever be realized — or, indeed, that someone apparently very healthy will not succumb to a major disease — then what are our options for intervention? How can we act to prevent ill health and promote a healthier population in an economically as well as medically viable way if we have no certainties with which to work?

A way of approaching and understanding this problem is suggested by Durkheim's 1897 sociological treatise on suicide.[33] Durkheim observes that suicide, although seemingly the most personal, private and individual of acts, cannot be predicted or fully explained at the individual level because, as with epigenetic responses, it is an act triggered by a contingent event acting on a particular propensity and whose occurrence and effect cannot be delimited in advance. Instead, there are supra-individual social forces at play so that, perhaps counterintuitively, at the level of the population in general there *is* regularity and predictability for we can, after all, calculate such things as suicide rates, birth rates, and crime rates. We can compare trends across time and plot differences between societies, and from these demographic indices we can extract information on societal characteristics that may explain the occurrence of these factors and suggest avenues of intervention.

This is a generalizable, calculable — and yet nonetheless mysterious[b] — feature of society that Foucault named biopolitics.[35] Biopolitics is not concerned with individual bodies but with the governance of populations. Foucault's elaborations of the concept articulate the same concerns as Durkheim and thus also address the problematic encountered in relation to the unpredictable nature of individual epigenetic responses to environmental exposures. Thus biopolitics is, contends Rose,[36] a politics of risk management.

Biopolitics is the consideration of the population simultaneously as a political, scientific and economic problem and as a biological

---

[b] For a further explication of the enigmatic phenomenon of cultural calculability, see Kirby [34] on the "unreasonable effectiveness of mathematics".

problem.[35] The phenomena with which biopolitics is concerned are, Foucault says, "aleatory" at the individual level but show calculable constants at the population level over time. The aim of biopolitics is to achieve overall states of regularity, and to realize this it attempts to forecast and anticipate, and so compensate for any randomness as it employs a continuous scientific power to reduce mortality and make the population healthier in a socially, politically and economically optimal manner.

Biopolitics is intimately linked with the operations of normalization, the processes of the continuous regulation of life, of making life and its context suitable for each other. It is the domain of value and utility, of measurement, appraisal and hierarchization.[37] Normalization is, however, neither good or bad, emancipatory or repressive, in itself. As with other processes discussed here, each specific instance of its operation must be evaluated in context. And it must be emphasized that to live in society is to be subject to continuous normalizing imperatives: it is, again, a process of making us and our society fitted to each other.[8] The biopolitical management of populations does not, however, necessarily mean that all benefit because, as Foucault contends, it is concerned with such issues as economic optimization and is also a mechanism of hierarchization. It does not automatically reduce inequalities and could just as easily further entrench them if that achieved the aim of regularity. Its concern is a general, rather than an individual, improvement and some may be deemed expendable as the cost of overall betterment.

Nor is biopolitics a process alien to biology or one that is imposed on it, for we could consider that biology already normalizes itself through its epigenetic responsiveness to its environment. As Kuzawa and Thayer,[19] Champagne[11] and Burdge and Lillycrop[7] explain, epigenetics studies the continuing adjustment and adaptability of an organism to its milieu, a construction of environment and biology as suitable to each other. And again, recalling Sanabria's musings on fat cells and junk food that suggest our involvement in the constitution of our own illness, this adaptation may not necessarily be what we think of as beneficial.

The Future: Narratives of Intervention and Responsibility

We are gaining a more in-depth understanding of the biological processes of epigenetics and in so doing achieving greater insights into disease and health. Bodily plasticity promises that we may be able to intervene by utilizing the body's own epigenetic operations to bring about change. Our growing knowledge may allow us to better identify, read and understand epigenetic biomarkers and refine our ability to identify people at risk, to better design or tailor drug treatments, and determine the most effective windows of time — in both personal and economic terms — for intervention.[18] However, these abilities to better manage epigenetic risk or susceptibility to disease remain at the population level and are not personal. As discussed, epigenetics is a non-certain and contingent process as there are a multitude of possible permutations in the complexity of any given individual-environment encounter and these risks cannot be calculated with any certainty beforehand.

The normalizing and generalizing operations of biopolitics thus necessitate a social conversation between government, population and agencies if interventions are to be successfully implemented. There needs to be a discussion about which methods or outcomes are deemed acceptable to members of *that* society at *that* time, because it is not feasible to conduct any such programs at the individual level. Because biopolitics treats population risk, not individual risk, it will not necessarily accommodate all interests. At the social level, this has to be weighed up in line with the ethos of the particular society, and the calculations and ordering of priorities needs to fit with this. An intervention is a political and economic program as much as it is a medical one as it is also concerned with what resources the society is prepared to expend. It is also a marketing issue. There has to be a narrative to explain and justify the intervention. As Stelmach and Nerlich[1] contend, our narrative of epigenetics shapes both the society and the social and political responsiveness to the narrative. It has to deal with questions such as what the particular society wants, how it sees itself, what its values and priorities are, and what costs — both monetary and in terms of people not treated — it thinks justifiable.

At this point it is relevant to note that there is nothing in these understandings of epigenetic possibility or biopolitics that deem them solely as forces for the improvement of society and the lives of the population. They are forces that could be manipulated in either a positive or a negative way, and the scenarios that play out may well be like those recounted in many dystopian imaginings of future societies. Or as a salutary lesson, we could look back at programs instituted under rubrics such as benevolent paternalism — programs such as the eugenic assimilation of indigenous peoples or the adopting out of the babies of unmarried mothers — to see how harshly history has judged what, at the time, were considered by their proponents to be largely positive interventions that were deemed to be "for the best". We can have no guarantee that any scheme instituted now will not suffer this same fate.

Epigenetic responsibility is political and not individual; it is a forward-looking social responsibility for change, not a backward-looking assignation of individual blame,[38] because, as I have discussed, epigenetic operations contest neoliberal or dualist notions of an individual's control over their life's outcomes. We cannot locate the point of individual agency because of the confounding spatio-temporal web of complexity that confronts us in such investigations. Epigenetic responsibility also needs to be long-term and sustained if, as Kuzawa and Thayer[19] suggest, epigenetic processes screen out short-term or intermittent changes.

The promise of epigenetics is twofold, although maybe distant. As well as allowing us a greater knowledge of the complexity of life and a more in-depth understanding of how we might ameliorate disease development and the consequent effects of illness, it also offers the prospect of reshaping society and politics. Epigenetic narratives could be seen as what Rose[27] terms a "philosophy of life", where our sense of how we should live is tangled with what we think we are as living beings. What epigenetics offers, if we have the social and political will to accept, is that our health problems may be more amenable to resolution if we move from the short-term thinking and political opportunism that seems to have failed us to a long-term intergenerational outlook. A reconceptualization of the individual, moving from a notion of the individual as the self-contained seat of agency to a more implicated view of ourselves as wholly enmeshed in our context may also shift responsibility away from concern just for ourselves, now, to also include taking a temporal responsibility for our environments and others. Because, as epigenetics entwines us with environment, politics and society, we may begin to see ourselves *as* environmental expression.

References

1. A Stelmach, B Nerlich. (2015) Metaphors in search of a target: the curious case of epigenetics. *New Genetics and Society* **34**(2):196–218.
2. M Meloni, G Testa. (2014) Scrutinizing the epigenetics revolution. *Bio Societies* **9**:431–456.
3. M Pickersgill *et al*. (2013) Mapping the new molecular landscape: social dimensions of epigenetics. *New Genetics and Society* **32**(4):429–447.
4. MR Waggoner, T Uller. (2015) Epigenetic determinism in science and society. *New Genetics and Society* **34**(2):177–195.
5. ME Pembrey *et al*. (2006) Sex-specific, male-line transgenerational responses in humans. *European Journal of Human Genetics* **14**: 159–166.
6. M Meloni. (2014) How biology became social, and what it means for social theory. *The Sociological Review,* DOI:10.1111/1467-954X.12151.
7. GC Burdge, KA Lillycrop. (2010) Nutrition, epigenetics, and developmental plasticity: implications for understanding human disease. *Annual Review of Nutrition* **30**:315–339.
8. N Rose. (2007) *The Politics of Life Itself: Biomedicine, Power and Subjectivity in the Twenty-First Century.* Princeton: Princeton University Press.
9. LA Joss-Moore, RH Lane. (2012) Epigenetics and the developmental origins of disease: the key to unlocking the door of personalized medicine. *Epigenomics* **4**(5):471–473.

10. MJ Meaney. (2010) Epigenetics and the biological definition of gene x environment interactions. *Child Development* **81**(1):41–79.
11. FA Champagne. (2010) Epigenetic influence of social experiences across the lifespan. *Developmental Psychobiology* **52**:299–311.
12. CW Kuzawa, E Sweet. (2009) Epigenetics and the embodiment of race: developmental origins of us racial disparities in cardiovascular health. *American Journal of Human Biology* **21**:2–15.
13. B Modin et al. (2008) The contribution of parental and grandparental childhood social disadvantage to circulatory disease diagnosis in young Swedish men. *Social Science & Medicine* **66**:822–834.
14. R Lund et al. (2006) Influence of marital history over two and three generations on early death. A longitudinal study of Danish men born in 1953. *Journal of Epidemiology and Community Health* **60**(6):496–501.
15. JS Alasaari et al. (2012) Environmental stress affects DNA methylation of a CpG rich promoter region of serotonin transporter gene in a nurse cohort. *PLoS One* **7**(9):e45813.
16. T Rönn, C Ling. (2013) Effect of exercise on DNA methylation and metabolism in human adipose tissue and skeletal muscle. *Epigenomics* **5**(6):603–605.
17. E Unternaehrer et al. (2012) Dynamic changes in DNA methylation of stress-associated genes (OXTR, BDNF) after acute psychosocial stress. *Translational Psychiatry* **2**(August):e150 doi:10.1038/tp.2012.77.
18. A Hoffmann, D Spengler. (2012) The lasting legacy of social stress on the epigenome of the hypothalamic–pituitary–adrenal axis. *Epigenomics* **4**(4):431–444.
19. CW Kuzawa, ZM Thayer. (2011) Timescales of human adaptation: the role of epigenetic processes. *Epigenomics* **3**(3):221–234.
20. G Nicolosi, G Ruivenkamp. (2012) The epigenetic turn: some notes about the epistemological change of perspective in biosciences. *Medicine Health Care and Philosophy* **15**:309–319.
21. H Landecker. (2011) Food as exposure: nutritional epigenetics and the new metabolism. *BioSocieties* **6**(2):167–194.
22. G Kaati, LO Bygren, S Edvinsson. (2002) Cardiovascular and diabetes mortality determined by nutrition during parents' and grandparents' slow growth period. *European Journal of Human Genetics* **10**:682–688.
23. KS Bishop, LR Ferguson. (2015) The interaction between epigenetics, nutrition and the development of cancer. *Nutrients* **7**:922–947.

24. J Wang et al. (2012) Nutrition, epigenetics, and metabolic syndrome. *Antioxidants & Redox Signaling* **17**(2):282–301.
25. S Luo et al. (2015) Differential effects of fructose versus glucose on brain and appetitive responses to food cues and decisions for food rewards. *Proceedings of the National Academy of Sciences*, www.pnas.org/cgi/doi/10.1073/pnas.1503358112.
26. E Sanabria. (2015) Sensorial pedagogies, hungry fat cells and the limits of nutritional health education. *BioSocieties* **10**:125–142.
27. N Rose. (2013) The human sciences in a biological age. *Theory, Culture & Society*, **30**(1):3–34.
28. Soylent. (2015) https://www.soylent.com/. Accessed 5 June 2015.
29. P Hinton. (2013) The quantum dance and the world's 'extraordinary liveliness': refiguring corporeal ethics in Karen Barad's agential realism. *Somatechnics* **3**(1):169–189.
30. F Chiew. (2014) Posthuman ethics with Cary Wolfe and Karen Barad: animal compassion as trans-species entanglement. *Theory, Culture & Society* **31**(4):51–69.
31. M Jamieson. (2015) The politics of immunity: reading Cohen through Canguilhem and new materialism. *Body & Society*, DOI: 10.1177/1357034X14551843.
32. V Pitts-Taylor. (2010) The plastic brain: neoliberalism and the neuronal self. *Health*, **14**(6):635–652.
33. E Durkheim. (1970) The Social element of suicide, in *Suicide: A Study in Sociology*. London: Routledge & Kegan Paul:297–325.
34. V Kirby. (2011) Enumerating language: 'the unreasonable effectiveness of mathematics', in *Quantum Anthropologies: Life at Large*. Durham & London: Duke University Press:49–67.
35. M Foucault. (2004) 17 March 1976, in *Society Must Be Defended: Lectures at the Collège de France, 1975–76*, M Bertani and A Fontana, Editors. London: Penguin Books:238–264.
36. N Rose. (2001) The Politics of Life Itself. *Theory, Culture & Society* **18**(6):1–30.
37. M Foucault. (1998) Right of death and power over life, in *The Will to Knowledge: The History of Sexuality Volume 1*. London: Penguin: 133–159.
38. M Hedlund. (2012) Epigenetic responsibility. *Medicine Studies* **3**: 171–183.

Index

acetylation, 12, 33
adaptation, 178, 180, 189
    local, 177, 188
adipocyte, 33
agency, 206, 209, 210, 212, 216, 217
age-related frailty, 109
aging, 182
    hallmark, 104, 112
    phenotype, 103
Agouti gene (A^{vy}/a allele), 137, 138, 162
animal models, 42, 45, 46, 49, 51, 54
antioxidant vitamin, 32

base excision repair (BER), 105
biopolitics, 212–216
    biology, 214
butyrate, 74

carbohydrates, 180, 190, 192
carnivory, 192
cellular senescence, 104
childhood obesity, 160
choline, 29, 30
circadian rhythm, 93
    BMAL, 94
    chrononutrition, 97

CLOCK, 94
CRY, 94
CRY1-2, 94
REVERB, 94, 95
coronary artery disease, 25
Cryptochrome, 94
cytosine and guanine nucleotides linked by phosphate (CpG), 5, 137
    island, 109

deacetylase, 12
developmental mechanisms, 188
    developmental machinery, 184
diabetes, 25, 28, 178, 183
dietary methyl donor, 115
dietary pattern, 32, 107
dietary restriction, 106, 179, 190, 191
dietary supplementation, 210
dieting, 209
dispersal, 185, 188
disposable soma theory of aging, 105
DNA hydroxymethylation, 7
DNA methylation, 3–5, 9, 11, 14, 27, 29, 31, 32, 35, 36, 41, 73–86, 161, 163, 164

age, 111
clocks, 110
mitochondrial, 113–115
DNA methyltransferase (DNMT), 6, 115
DNMT1, 6
DNMT3A, 6
DNMT3B, 6
DNA repair endonuclease XPG (Gadd45a), 6
DNA repair gene, 106
DNA repair system, 105
docosahexaenoic acid (DHA), 75–78, 81–83
Dutch Hunger Winter, 26, 28

early conditions, 179, 181
early-life experiences, 184, 185
early-life nutrition, 180
long-term effects of, 180
early life origins of disease, 26
eicosapentaenoic acid (EPA), 76, 78, 82, 83
ELOVL2, 83
ELOVL-5, 83
environmental change, 188, 193
environmental predictability, 185
environmental stability, 187
frequency of change, 183
variable environment, 185
enzyme inhibitor, 115
epigallocatechin-3-gallate, 116
epigenetic marks, 186, 187
DNA methylation, 187
epigenetic modifications, 179

epigenetics, 2, 3, 25, 26, 31–33, 35, 36, 43–47, 49–54, 56–58, 147
biomarker, 160–162, 165, 167, 168
definition, 202, 203
maintenance system, 112
mitochondrial, 113
non-determinacy, 204, 207, 213
plasticity, 27
programming, 55
promise, 216
social aspects, 205
spatio-temporal aspects, 216
temporal responsibility, 217
theory of inheritance, 187
time-scale, 205, 206
epigenetic-environment entanglement, 203, 204, 206, 208, 211, 215
epigenome, 25, 37
evolutionary medicine, 184, 192

factor 2, 27
Fads1, 78, 83
Fads2, 77, 78, 81, 83, 163
famine, 187, 191
fetal programming, 27
folate, 29
folic acid, 29, 137, 139
food entrainable oscillator (FEO), 98
Foucault, Michel, 213, 214

genetic divergence, 177, 179, 188, 189, 193
genomic imprinting, 13, 14, 186
Geometric Framework, 179, 191, 192

glucocorticoid receptor (GR), 163
growth, 177, 180, 183–185, 190
    catch-up growth, 181, 182
    compensatory growth, 182
gut microbiota, 32

5 hydroxy methyl cytosine (5hmC), 7
H19 methylation, 34
heterochromatin, 12
high caloric diet, 141
high fat diet (HFD), 142
histone, 31, 42, 43, 45, 46, 48, 52, 55, 56
    acetylation, 10, 16, 74, 75
    acetyltransferase, 10
    deacetylase (HDAC), 10, 75, 84
    methylation, 10
    modification, 3, 8, 9, 12, 15, 41
    phosphorylation, 10
hormones, 185
human evolution, 178
    evolutionary history, 178
    lactose tolerance, 188
hyperglycemia, 29

IGF2, 33
IGF2/H19, 34
individual, 201
information, 184–187
    cues, 181, 183–186, 188, 193
    informational, 185
interdisciplinarity, 201

life history, 179, 184, 190, 193
    strategies, 193

lifespan, 178–180, 183, 189, 190
α-linolenic acid (ALNA), 77, 78
linoleic acid (LA), 77
long interspersed element 1 (LINE1) methylation, 108, 109

macronutrients, 190, 191
    carbohydrates, 178
    fat, 178, 191, 192
    protein, 178
malnutrition, 143
matching, 181
material-social entanglement nutrition, 210
maternal effects, 187
    maternal nutrition effects, 181
Mediterranean diet, 32, 107
metabolic syndrome, 209, 212
methyl binding domain protein (MBD)2b, 6
methyl CpG binding protein 1, 5
methyl CpG binding protein 2 (MeCP2), 12
5 methyl cytosine (5mC), 4, 6
methyl deficient diet, 139
methyl donor, 27, 29
methyl supplemented (MS), 137
methylation, 33, 42, 44–57
methyltransferase, 10
microbiota, 189
    bacterial symbiont, 188, 189
    microbiome, 188
microRNA (miRNA), 33, 41–45, 49, 52, 55–56
milieu, 203, 204, 211, 214
Millennium Development Goals, 159, 170

mismatch (developmental or evolutionary), 178, 179, 186, 191, 192
mismatched, 183, 184, 191
mitochondrial genome, 112
monounsaturated fatty acids, 32
monozygotic (MZ) twins, 108
multigenerational, 135, 139, 147

non-coding RNA (ncRNA), 3, 13, 161
non-communicable disease (NCD), 25, 159, 162, 165, 167
non-CpG methylation, 5
non-genetic inheritance, 189
normalization, 214
NR1D, 94, 95
nucleotide excision repair (NER), 105
nutrient
    material-social entanglement, 208
    sensing, 104
nutrition
    behavior modification, 208
    material-social entanglement, 207, 208, 210

obesity, 25, 28, 34, 36, 106, 113, 178, 179, 183, 189, 191, 192
obesity-associated inflammation, 107
oleic acid, 75, 76
omnivory, 192
    omnivorous, 178
over-nutrition, 28

parent-offspring conflict, 186
paternal, 33

PER, 94
PER1-3, 94
periconceptional environment, 25
PERIOD, 94
peripheral clock, 95, 97
peroxisome proliferator activated receptor alpha (PPARα), 163
phenotypic inertia, 206
phosphorylation, 12
physiological damage, 183
    cellular damage, 182
    oxidative stress, 183
phytochemicals, 33
plasticity, 181, 201, 205, 215
    adaptive, 183, 184, 186, 188
policy, 202
polycomb repressive complex, 11
prebiotics, 32
predictive adaptive response, 181
preeclampsia, 31
pregnancy, 77
programming, 44, 46, 47, 49, 51, 53, 54, 56–58
proopiomelanocortin (POMC), 163
protein, 180, 190–192
protein restricted diet, 145
proxy tissue, 164
public health policy, 170

resources, 180, 182–186, 189, 190, 193
    resource allocation, 190
retinoid X receptor, 31, 34
rodent model, 43
Rose, Nikolas, 204, 211

S-adenosylmethionine, 30
selection, 178
short chain fatty acids, 32

silver spoon, 180, 181
single-cell level, 109, 119, 120
suprachiasmatic nuclei (SCN), 95
Sustainable Development Goals, 170

Ten-eleven translocation (TET) protein, 6–8, 16, 85, 113
The Roadmap Epigenomics Project, 165
thrifty genotype, 191
transcription, 4
    factor, 161
transcriptome, 75
transgenerational, 36, 135, 141, 143, 147

transgenerational effects, 179, 186–188
    maternal effects, 185
    maternal nutritional effects, 185

Vitamin B12, 141
Vitamin D, 31, 146

Western diet, 163
Western dietary pattern, 32
white adipose tissue (WAT), 97
World Health Organization (WHO), 159

X inactivation, 15